GEOGRAPHY OF THE SOUTH ASIAN SUBCONTINENT

A Critical Approach

About the Author

Aijazuddin Ahmad was a well-known geographer and academician. He taught at the Jawaharlal Nehru University, New Delhi, for twenty-five years. Earlier, he taught at the Aligarh Muslim University from 1962 to 1972. His areas of interest include historical, social, and regional geography. His research was particularly focused on the study of tribe, caste, language and minority groups.

Prof. Ahmad was involved in several interdisciplinary exercises at the behest of National Institute of Educational Planning and Administration (NIEPA), National Open School (NOS), the National Council of Educational Research and Training (NCERT), CSIR and ICSSR.

He has edited several Indian geographical journals and has held positions, like President of the National Association of Geographers, India (1997-98), Member of International Geographical Union (IGU), Commission on Population Geography (1990-94), and Chairperson of the National Committee of the Indian National Science Academy for IGU (1993-97).

GEOGRAPHY OF THE SOUTH ASIAN SUBCONTINENT

A Critical Approach

AIJAZUDDIN AHMAD
Formerly Professor
Centre for the Study of Regional Development
Jawaharlal Nehru University

CONCEPT PUBLISHING COMPANY, NEW DELHI-110059

Cataloging in Publication Data—DK
Courtesy: D.K. Agencies (P) Ltd. <docinfo@dkagencies.com>
Ahmad, Aijazuddin, 1932-
Geography of the South Asian subcontinent : a critical approach / Aijazuddin Ahmad. — 1st ed.
p. cm.
Includes bibliographical references (p.).
Includes index.
ISBN 13: 9788180695681
ISBN 10: 8180695689

1. South Asia—Historical geography. 2. South Asia—Social conditions. 3. South Asia—Politics and government. 4. Economic development—South Asia. I. Title.

DDC 911.54 22

ISBN-13: 978-81-8069-568-1 (HB)
ISBN-10: 81-8069-568-9 (HB)

First Edition 2009

Published and Printed by
Ashok Kumar Mittal
Concept Publishing Company
A/15-16, Commercial Block, Mohan Garden,
New Delhi-110059 (INDIA)
Phones: 91-11-25351460, 25351794
Fax: 91-11-25357109
Email : publishing@conceptpub.com
Website : www.conceptpub.com

In Loving Memory of My Parents

Buaa and Mian

Who Taught me the Meaning of Poverty
and the Art of Living in Poverty with Dignity

Nanak ne jis zameen par
Wehdat ka geet gaya
Mera watan wohi hai
Mera watan wohi hai

The land where Guru Nanak came
With his message of eternal unity
That is my homeland
That is my homeland

Allama Iqbal
Poet of the East

Preface

Geography of the South Asian Subcontinent: A Critical Approach, among other things, explores the possibility of reviving and rediscovering the unity of the subcontinent at a higher level of generalisation. The subcontinent and its people constitute a single ethnic formation. It is not a territorial unit only, although it is internally divided into several countries on the criteria of ethnicity, religion, faith, language, geographical location and shared history. The subcontinent is in actual effect a unity of the first order. Is it a viable idea that we are one people divided into several nations? One may posit that the unity of the region is implicit as well as explicit. Its boundaries have been demarcated by geography and sustained by history. Sometimes it is suggested that the idea of the subcontinent can be revived and implemented by unifying the whole territory from the northwestern border of Afghanistan (Bamiyan and Mazar-e-Sharif) to Manipur and Mizoram in the northeast and from Hunza, Gilgit and Baltistan in the north to the tip of the Indian Peninsula and the neighbouring islands of Sri Lanka and Lakshadweep in the south. A confederacy of this kind would make massive defence expenditure a redundancy if not an absolute absurdity.

My earlier book on *Social Geography* (Ahmad 1999) presented a comprehensive view of the elements of geography of India as we know it today. However, India today is a residual territory, *i.e.*, what is left of the subcontinent after excluding Afghanistan, Pakistan, Nepal, Bangladesh, Bhutan, and Sri Lanka. For certain reasons the book could not cover the entire subcontinent. Moreover, it left many questions unanswered. For example, the corridor role of Afghanistan in the process of peopling of the subcontinent could not be highlighted. Similarly, recent migrations within the subcontinent, such as those of Afghans and the Kashmiri Pundits, could not find a place. It is believed that these issues cannot be discussed in isolation.

It may be recapitulated that almost all incursions into India were made via Afghanistan. The region was recognised as the nerve-centre of the successive empires in South Asia. It is well known history that the streams of migration, which reached India and affected the culture and the people, passed through Afghanistan. The Afghan territory served as a doorstep for the travellers who intended to visit South Asia. It worked like a funnel for the movement of armies of invaders or bands of nomads. It is a truism that any attempt at understanding the society, its social history, and the social antecedents of the people would be incomplete without relating it to the history of Afghanistan and its space relations with Central Asia (Turkmenistan, Uzbekistan, Kazakhstan, Tajikistan, Kyrgyzstan) and Iran and with other countries lying beyond. It is necessary to know Afghanistan in order to know the subcontinent. For long periods of history the Afghan territory was ruled from the Punjab. On the other hand, political power based in Afghanistan ruled over Punjab and parts of Sind. There were episodes when emperors, such as Maurya, Ashoka, Harshavardhana, Alla-uddin Khilji and Akbar, to mention only a few, considered Afghanistan as an integral part of their respective empires. Similarly, several wars between the Afghans and the British, based in India were fought to expand the British Empire over the traditional Afghan territory. Therefore, it is necessary to look back into the past in order to know the present. *Social Geography* focused mainly on India. Obviously, there was little scope for making lengthy references to the social scene in other countries of the region. Geographers, archaeologists and historians have laid stress on a certain number of factors, which supposedly influenced the region as a whole. India and the subcontinent are inseparable. For example, each country passed through the same phases of social and political history, by and large similar in evolution of polity and developmental experience. People of the subcontinent have a shared history, particularly the history of British Raj in the eighteenth and nineteenth centuries. The Raj bound them together and left imprints on all parts of the subcontinent from Afghanistan to Sri Lanka and all other spheres of life of the people of the subcontinent.

This book traces the present problems of the people in the region and evaluates the ground realities in order to rebuild confidence. It is hoped that the South Asian nations would rise to the occasion. Past has to be forgotten to allow peace a chance. Besides, the book also evaluates the present predicament of the leaders of the South Asian nations. It traces the history, more importantly social history, discusses the ramifications of the new arrangements, in the perspective of the evolution of polity at the newer and higher economic plane in which the more intensive and varied economic and cultural exchanges would take place between and among the South Asian nations. The book makes a case for a new union of the nations of the subcontinent on the model of the European Union. It may be assumed that the basic idea is transcendental to the existing political organisation of the region. This is the only way to ensure peaceful relations for sustainable human development. The basic purpose is to remind the governments and the people of the region to bury the history of discord in order to avoid bloodshed in the name of religion, nationality, territorial integrity, caste, and ethno-lingual identity.

Perhaps the concept of nationhood in the new millennium would be based on entirely new criteria. The new criteria for state formation within the union would not help the war-hungry politicians on either side of the newly created borders and the barbed wire fences. In fact, there would be no place for the politicians and their military establishments, including the inter-continental ballistic missiles and the nukes, if this scheme of reunification is implemented in letter and spirit. Of course, there will be a joint armed force at the subcontinent level to defend the interests of the people of the region as a whole, both severally and collectively.

Admittedly, the proposed idea of unification looks utopia. Only time will tell how far we are from our goal, if not acceptable, time will throw it in the dustbin of history, only to be picked up by the next generation South Asians, or the generations next to them.

Palam Vihar, Gurgaon
January 2006

Aijazuddin Ahmad

Contents

List of Tables

List of Boxes

List of Maps

List of Abbreviations

ABR	:	Assam-Bengal Railway
ADB	:	Asian Development Bank
AIADMK	:	All India Anna Dravida Munnetra Kazhagam
AJGAR	:	Ahir, Jat, Gujar, Rajput
ASI	:	Archaeological Survey of India
BCIR	:	Bombay and Central Indian Railway
BNR	:	Bengal-Nagpur Railway
CIDA	:	Canadian International Development Agency
EIR	:	East India Railway
EU	:	European Union
GDP	:	Gross Domestic Product
GIPR	:	Great Indian Peninsular Railway
GNP	:	Gross National Product
GSI	:	Geological Survey of India
GTR	:	Grand Trunk Road
IBM	:	Inter Ballistic Missile
IMF	:	International Monetary Fund
IPC	:	Indian Penal Code
LOC	:	Line of Control
MSMR	:	Madras and Southern Maratha Railway
NATO	:	North Atlantic Treaty Organisation
NPA	:	National Progressive Alliance
NWFP	:	North Western Frontier Province
NWR	:	North Western Railway
OTR	:	Oudh and Tirhut Railway
PAC	:	Provincial Armed Constabulary
RKR	:	Rohilkhand and Kumaon Railway
PR	:	Permanent Residence
SOI	:	Survey of India
TDB	:	Tribal Development Bank
UK	:	United Kingdom
UN	:	United Nations
UNDP	:	United Nations Development Programme
USA	:	United States of America
USAID	:	United States Agency for International Development

Glossary

Akhand Bharata	Indivisible territory usually referred to as Bharata or Bharat Varsha
Anusuchit Jati	Castes mentioned in a certain schedule of the Constitution of India
Berang	Tasteless, something which fails to make a mark
Bhadralog	A Bengali word for the rich upper classes of people
Bhandaras	Big conclaves of caste groups held in specific locations to consult the caste elites in the decision-making process
Bhikshus	Buddhist sadhus, monks
Biryani	A dish prepared with rice and meat, a Mughal delicacy
Chilghoza	A variety of Himalayan pine, its seed or kernel (*pinus geradiana*)
Darbar	A royal court, hall of audience
Dalit	Persecuted masses, generally scheduled castes
Dhamma	A variant of dharma which means the responsibility of the code of conduct of the ruler for sustaining the state
Doab	Land between the two confluent rivers, e.g., Ganga-Yamuna doab
Farsi	Persian, language spoken in Iran or by people of Persia
Firman	A Royal decree or charter
Gharanas	A school of Hindustani classical music
Ghazals	The origins of ghazal lie in the Arabic and Persian language poetry, Urdu adopted it as a form of poetry
Istan	Place of residence or abode
Jagah	Terrestrial space, a particular place, locality or a location
Jamhuriya	A republic
Janapadas	People's republics
Jataka	An anthology of Buddhist stories

Jehadi	A crusader or someone committed to the cause of Islam
Jhum	Shifting agriculture, a term used generally in the Northeastern parts of India, a system of agriculture also known as 'slash and burn' agriculture
Jirga	Clan or a sub-clan
Kaddiyo	A Brij Bhasha variant of action already finished, an action following the orders of a senior member
Kateeb	Headman, an administrative functionary in the Maldives
Khandsari	A cottage industry engaged in processing the sugarcane juice in India, its by-products include sugar, jaggery and sugar powder
Khas tatti	A curtain or a hanging made of the wild grass known as khas (the grass/*Andropogon muraticum*)
Kulfi	A local name for an ice cream prepared with full-cream milk and sugar, a desert
Lagaan/malguzari	Land revenue, land tax either paid to the zamindar or deposited directly in the government treasury
Lashkari	Belonging to or of the army
Li	A Chinese measure of length, about one-sixth of the British mile
Lok Sabha	Lower body of the Indian Parliament roughly comparable to the Lower House of the British Parliament
Madrasa	A primary school for instruction in Islamic religion/theology and a code of conduct
Madhyadesa	Midland, central parts of the North Indian Plain
Maktabs	A lower level school for imparting Arabic, Urdu education, particularly knowledge of Islamic theology; a centre for religious instruction
Mahabharata	Name of an ancient Indian epic
Maharaja	Someone holding the powers of the *raja*
Mawara-un-nahar	Land beyond the *nahar* (Oxus) in Central Asia; Turan or Eastern Turkistan
Munsif	A lower order judicial magistrate
Murgh-e-musallam	A dish of roasted chicken, a Mughal delicacy
Nawab	A feudal lord, generally in use among the Muslim middle class
Nava-khanda	Nine divisions, regions of the subcontinent in ancient India
Nirvana	Salvation, a term generally used to refer to ultimate release of the spirit from the body, a Buddhist gospel

Pahari	A hillock or a hilly tract of land, something mountainous, a Hindi dialect spoken in hilly areas
Pak	Pure, clean
Panchsheel	Five cardinal principles of Buddhism, usually recommended for global fraternity and peace; Nehru adopted the term as a guiding principle for international relations; a major plank of his foreign policy
Pathshala	Lower-level school especially a primary school for imparting the education in Sanskrit
Patwari	A keeper of the land records at the village level
Perestroika	A Russian language word meaning opening up or reconstruction
Prakrit	A local dialect; a form of spoken language; a variant of the original Sanskrit, for example, Pali
Puranas	A term generally used to refer to a body of floating knowledge, both religious and secular in ancient and early medieval period
Raja	A king or a ruler
Raj dharma	Law to govern the functioning of a state, *rajya* and the *raja*
Raj Purohit	A clan/family priest
Rajya Sabha	The Upper House of the Indian Parliament comprising indirectly elected as well as nominated members, with each state having its fixed quota; the term lasts for two years only
Ramayana	An Indian epic describing the story of Lord Rama
Registan	A dry region or a desert, sandy or rocky
Sal	Name of a tree (*shorea robusta*) found in the sub-Himalayan region in north India and on the eastern flanks of the Indian plateau
Samaj	Society, a social formation; a social collectivity
Saptasindhava	Land of the seven rivers of northwestern India belonging to the Indus system, e.g., Sutlej, Beas, Ravi, Chenab, Jhelum, Indus itself and the legendary Saraswati
Shudh	Pure, original, undiluted
Shudhikaran	To purify or to cleanse; a basic Hindu project
Stadia	A Greek measure of length equal to about 85 metres
Sthal	Place, location, locality, place of abode
Sthan	Place, area, region
Subedar	Provincial governor, particularly during the Mughal period
Sufi	Saints belonging to a mystic tradition promoting secularism and inter-community harmony

Sutras	A short rule or aphorism generally used in Sanskrit manuals of instruction
Swaraj	Self-rule, a term, which was adopted as a slogan of the people of the Indian subcontinent during the independence struggle
Tehsil	A land revenue sub-division within a district; one or several tehsils may constitute a district
Tehsildar	A lower-order judicial official or a magistrate with his jurisdiction lying within the tehsil
Terai	A wet region generally referred to the strip of the North Indian Plain
Thug	A member of a secret gang of highway robbers, a cheat and a fraudulent individual, particularly mentioned in the early British rule in the subcontinent
Tika or tilak	A paste of red ochre applied on the forehead of the king at the time of his coronation
Umr'aa	A body of ruling elites who functioned as an overseer guiding the king, particularly during the days of Delhi Sultanate and dynastic rule of the Mughals
Varna	Originally means colour of the skin; the original four-fold Varna system, the basis of further caste ramifications in ancient India

1

Introduction

This book is a journey of exploration. The author believes that it is necessary to explore the multiple dimensions of the society and culture of the South Asian region. The subcontinent has passed through a long history of human settlement and regional cultural differentiation. While the parts of the subcontinent have been ruled by dominant groups, both insiders and outsiders, it is argued that geographical location, nature of the terrain and climatic factors have made a significant contribution. This has led to the emergence of the state. In certain regions of the subcontinent states did not emerge for long periods of time. The major determining factors were the location and the natural resource-potential. Accessibility by land and sea provided the basis on which some regions emerged earlier in political organisation while others lagged behind.

Purpose of the Book

The purpose of this book is to look back in time to find the roots and the social origins of our people. The idea is to explore the basis of our identity as Indians and as South Asians. An attempt has been made here to rediscover and restate our status as individuals and as a culture-group. In our view there is a need to rediscover the origins in the light of our past history without any prejudice towards the people and the different sections of the society. Social configuration suggests intermixing of ethno-lingual traits as moulded by the conditions of our existence. They represent different ethnic groups. The people of the subcontinent have passed through several phases in history. However, it is necessary to know that we live at different levels of consciousness of our status as a citizen of the country as well as the citizen of the subcontinent. Ethno-lingual identity is expressed in many ways. This identity is based on a set of factors, such as our personal names and social behaviour. The people are categorised like the Hindus, Muslims, Christians, Sikhs and the Buddhists. Categorisation has further promoted our knowledge of division on the basis of castes, sub-castes, tribes, dialects, languages and other cultural parameters. The fact is that with the changing times the dividing lines have not been rubbed off. It is generally understood that the divisive factors are a product of the system of rigid social hierarchy, which determines the status and social rank of the ordinary men and women of the subcontinent. The book makes an appeal to the reader to envision an India, which is a cumulative whole and functions as an integrating process. We are members of a community, which is indivisible. We are Indians first and then anything else. The people belong to several religious persuasions; geographical location, North Indians or South Indians; tribes; insiders and dwellers; hillsmen with clusters of population in villages spread over the river valleys, such as the Indus, Ganga, Brahmaputra, Godavari and the Kaveri. The population is composed of different ethno-lingual groups like the Gonds, Bhils, Santals, Nagas, Nishis and Bodos, to mention only a few. Then, there are regional and ethno-lingual categories like the Kashmiris, Biharis, Punjabis, Bengalis, Malayalis, Tamils, Kannadigas and others. Divisions are there in social stratification: high castes, low castes; high Brahmins, low Brahmins; untouchables, outcastes and so on. As already stated above, religious consciousness has created further divisions like the Hindus, Muslims, Christians, Sikhs, Buddhists and others. Many a time the geographical location leads to further divisions: North, South, East and West. Behavioural groups have emerged and the psychological problems express themselves in different ways. It is common knowledge that these divisions would linger on. The evidence suggests that inter-community relationship has been changing. Sometimes insiders and outsiders are differentiated. In some cases groups are perceived as foreign elements and condemned as outsiders. The psychological perception and expressions like 'we' and 'they' have contributed to an ethos, which does not match with the spirit of a liberal democracy. However, these divisions are skin-deep and should have been forgotten by now. A political culture has emerged which adopts religion, caste and tribal identities for electioneering and for capturing power. Electoral politics has glorified the old categories, like the scheduled castes, scheduled tribes, other backward castes and the untouchable groups, like the *dalits*. The masses are conscious of these divisions, but they lend themselves to be used as potential vote banks. During the election campaigns, political parties often talk of Muslim lobby, Hindu lobby, Sikh lobby, etc. or Muslim vote-bank and Christian vote-bank. However, people are not generally aware of the social boundary lines. The modernisation process during the last phase of the colonial history, as well as in the contemporary period since independence, has given a new lease of life to the discriminatory philosophies and the native tradition. There is a correspondence between the political forces and the caste-based or religion-based formations. It has contributed to a further hardening of attitudes in inter-community relations. It is generally known that the subcontinent passed through a phase of turmoil in the pre-independence period. The social mechanics and the policy directives neatly conform to each other. More than half a century of independence has brought about a dramatic change in peoples' behaviour, but education and social development have failed to create the vision of an all-India unity. The nations of the subcontinent have been behaving differently in different parts of the subcontinent. Policies have been pursued to further strengthen the divisions rather than rejecting them as archaic and redundant.

We have argued that these differences have percolated down the levels of social hierarchy. The nations of the subcontinent have worked out an agenda and a political strategy, which damages the vision of an all-India unified state. Geographical distribution of this tradition has captured the mind. New terminology has been coined in a way that destroys the centuries of living together and inter-community harmony. Social determinism has created a mythology of superiority and social discrimination based on it. The state today operates through an administrative machinery where status is a function of social rank. For example, if the members of the low castes, the *dalits*, join the administrative service or the law-and-order maintaining forces, like the police, in actual terms they are not treated equally. The age-old discriminatory system has been strengthened further in consonance with the principles of traditional rank and hierarchy.

This book examines the roots of our social discriminatory behaviour and the people's consciousness of the rank and hierarchy. The masses are not conscious of the roots, and therefore, they suffer because of the lack of understanding. Democracy has further widened the cleavage between people and the chasm between the old and the new has been further widened. Today the people of the South Asian subcontinent have reached the stage of no return. The objective of an egalitarian society is still out of sight. This has happened because only a small section of population has been exposed to the processes of modernisation and the spread of modern education. Heredity is set apart by the old ideologies which are definitely divisive. We declare our commitment to the values of democracy, secularism and social justice. The common people of India, Pakistan and Bangladesh are generally bypassed by the political elites and further marginalised by the developmental process. Half a century of democratic functioning has not been able to relieve the people from the stranglehold of tradition. The classical view that it is the birth that determines status or rank in life has survived the centuries of reformist movements. This book tries to expose the people of the subcontinent to the social reality of the day. More than five decades ago the country was divided in the name of religion. At another stage thirty-five years ago, another partition separated Bengali Muslims who were liberated from the stranglehold of the Pakistani establishment in the erstwhile East Pakistan. The *dalits* in India have been waging a war of liberation without much success. India is the largest country in South Asia with its awe-inspiring volume of population and the size of geographical area. The civil society has not been able to reject the philosophies of social determinism and replace them by egalitarianism. The law of the land is that the stronger should control the weaker and in the process the weaker shall not benefit from the policies aimed at achieving overall development, both economic and social. These are no ordinary questions.

It is hoped that the ideas advanced in this book will contribute to a new consciousness, which would reject the age-old discriminatory institutions and practices. The book, it is hoped, will initiate a process for achieving a holistic development and for resolving the crisis of our life and times. It may instil a new hope and self-confidence in the minds of the people. The churning of these ideas may lead to a new social man, committed to the cause of social justice. It is also hoped that the people of the subcontinent would reach a new and a higher level of cultural synthesis, rising above the political divisions, states and the political boundaries. The message is that we are one people and all divisions are superficial. But we cannot forget that the boundaries of the states are not going to be erased and will continue without any break from the past. The nations are bound to defend every inch of the soil as their own and will do everything possible to defend the boundaries. We are capable of rising above the rhetoric of the discriminatory viewpoints.

The Post-Modernist View

The intellectual roots of this book lie in the philosophy of post-modernism. The process of globalisation has brought the people face to face to the new impulses of change. In fact, globalisation has weakened particularisation in all parts of the subcontinent. This has by and large questioned the older views. It appears that the age of trans-nationalism has weakened the concept of a nation state. We live in a global village where everybody is connected with everybody. The electronic devices of inter-personal communication and conferencing have changed the situation drastically. In fact, the transfer of new knowledge takes place in seconds without making any effort. The concept of nation-state is undergoing a change in people's perceptions. Perhaps the people are not prepared to create a new ethos of living in a different world without boundaries and frontiers. There are certain world realities, which cannot be ignored. One of the facets of this changing world is globalisation. In the context of this change continents appear like islands. We find little hurdles in the way of reaching others in any part of the world either by air, overland or the sea routes. An aeroplane can take us to a distant place in Greenland, or the base of the Elbruz Mountains in Russia in a few hours, not days. A telephonic call can connect the leaders of the most deprived nations in Africa or Oceania with the president of the mightiest power on the planet earth—the United States of America—in a split second. The leaders, captains of business and industry, elites, commoners, tourists and students execute their travel plans on the mobile phone, booking their accommodation in class hotels and confirming their aeroplane seats. For many people, overland travel is no longer a desirable proposition. Distances are shortened by money. Sponsorship for academic work abroad is available with a little effort.[1]

When it takes just ten hours to reach London, Paris or Amsterdam from New Delhi, the geographers' concept of regions and regionalism needs to be revised thoroughly. The international market of goods and services has expanded to such an extent that small economies and small countries have been willy-nilly integrated with the world market. The multi-national companies have expanded their catchment area from end to end, thus encompassing the whole of the planet earth.[2]

In so far as written or oral communication is concerned American English dominates the scene. It may be noted that all the glorious languages of the world have lost ground due to increasing globalisation of the English language. Similarly, the old concepts of geography based on the Hartshornian logic of spatial distribution and regional differentiation appear to have been eroded by the swollen river of the electronic media, such as the Internet. All these developments, it appears, have rendered geography an anachronism.[3] But geography is there and it is going to stay. The far-off locations as well as localities are there and they will continue to inspire new geographers to play on the new pitch of local area networking and the Internet. The pursuit of knowledge is an intellectual exercise. Computers give an encapsulated knowledge on diskettes and the Internet, but they cannot replace the human intellect. For geography, there will be enough room to tread upon revisited themes in the years to come. The geographers have explored and experienced the terrestrial space in all its dimensions. Now they should explore the unexplored on the cyber space. Little nations and little cultures are there and they need further exploration in order to organise them in the new models of unionisation and collectivisation. The European Union is the most recent example of this phenomenon.[4] It has been a story of success as well as failure. What Europe has done today will be replicated in Asia, Africa and Latin America, if not sooner than later. There are new frontiers to be reached and new bridges to be crossed.

A critical approach to the geography of the subcontinent is perhaps necessary at this stage of economic and human development.[5] The problems of today can perhaps be understood in the context of time. This implies that the past geographies have to be reconstructed systematically. Old arguments like the integrity of the state and the territorial sanctity were invented by the intellectual leaders of the age of modernism. Today, these ideas have lost much of their meaning. In fact, the contemporary trends in development show that trans-nationalism would prevail upon sectarianism. This is the logical conclusion one can draw from the prevailing situation in the international arena.[6] Nations of the South Asian subcontinent should learn to live with the new ideas of integration and cooperation in all spheres of life. Their future lies in cooperation and not in confrontation. Ideally, it would be better if people-to-people contacts displace the philosophy of confrontation at the political levels. But we are not able to throw the yoke of foreign interventions. However, small nations and little countries of the world have not reached that stage where boundaries can be so simply removed. In earlier writings we have noticed a desire to form a world government, which will finally subsume the entire world and the present countries would be like states or provinces. The substantive point is that only humanistic values can bridge the gap in understanding between the people and thus pave the way for the eventual unification of the world society into a single socio-political order: the order of capitalism. It is a harsh reality, but this is what the reality is. In fact, South Asia presents an interesting scene in so far as aid-giving and aid-receiving policies are concerned. India, for example, receives military weapons from Russia and other parts of the developed world. The same country, *i.e.*, India, supplies arms and financial aid to smaller countries like Nepal.

The South Asian subcontinent is often described as a museum of people, cultures and ways of living. Today, the population of the subcontinent largely consists of the sedentary peasant communities, who are only vaguely conscious of their diverse ethnic origins. The population also includes nomads and pastoralists, or pastoral nomads, who do not have a shared history. There are caste-groups who are basically unequal in their social rank and occupation. They are arranged along a well-defined vertical scale, which shows that the society is essentially hierarchised and stratified. Understandably, democracy in a hierarchical society is likely to be a mismatch. Let us face the facts. An overwhelming section of population consists of the low castes, some of them still perceived as untouchables or *dalits*[7]. In India all adult individuals have a right to vote. The *dalits* have been carrying on their shoulders a history of persecution and subjugation by the overlords of the quasi-feudal society. However, electoral politics, particularly in India has helped some of the lagging castes encash on their 'backwardness' and poverty. But, official claims notwithstanding, their backwardness to some extent has increased with time and continues unabated. The gap between the rich and the poor is widening, even though they are moving ahead at snails pace. The Constitution of India has mentioned several caste-groups as scheduled castes (*Anusuchit Jatis*) and some ethnic groups as scheduled tribes (*Anusuchit Janjatis*).[8] It may be noted that the Constitution of India has been amended time and again to accommodate the marginalised groups consisting of both castes and tribes. These remarks apply to the people of India more than to any other country of the subcontinent. New categories, such as the 'other backward castes' (OBCs) have been recognised and protected by the state. The population includes millions of tribes with diverse ethnic origins, ways of living and cultural characteristics. They profess different religions and communicate in different dialects. But in a liberal post-modern society, communities, castes and languages are not entered in the schedules of the constitution/statutes. They are free to choose their own path to development. This practice should be abandoned. One solution is that all the communities should be listed in the schedule in accordance with their proportionate share in the total population.

Democracy cannot breed discriminatory practices. The basic idea is that all people are equal and have the same rights to whichever strata they belong. However, the social system of the subcontinent allows discrimination in the name of caste, community, religious faith and social status.

The people of the subcontinent display a vast diversity in traditions of arts, crafts, legends, folk songs and the folk dances. Similarly, there are striking differences in their food habits and dress styles. This diversity has persisted over the millennia of shared history.[9] It may be noted that both Buddhism and Islam discarded preferential discrimination as a policy of the state. Islam rejected the ranking of the communities or individuals on the basis of caste, community, creed or class. But the Muslim society of the subcontinent could not reach this goal. The low castes and the community groups have sometimes organised themselves in political parties in order to struggle for their rights at the social as well as the political levels, e.g., Bahujan Samaj Party of India is an apt case. These groups somehow survived the onslaught of time, but could not relieve themselves from the clutches of tradition.[10] Even before independence there were parties, such as the Zamindar Party of the Punjab, representing a class of quasi-feudal lords. Regions identified on the criterion of ethno-lingual homogeneity by and large have retained their original cultural traits. It has been suggested that both history and geography are responsible for sustaining these diversities over time and across the regions. The older traits have survived despite the supposed road-rolling effect of developmental planning and the growing inter-regional exchange of goods and services. The far-flung parts of the subcontinent are now exposed to the compulsions of the international economic order.[11] How can globalisation help the fragile economies of the lagging regions of the subcontinent? It is observed that disparity owes its origin to the non-performance of the governments at the central or the state and the provincial levels. But this is not the whole truth.[12]

The Indian concept of backwardness often defies definition. Perhaps backwardness is rooted in the diversities in the people's struggle for survival, and in their slow response to the impulses of social and economic change. Generally, continuity and change chase each other. In many cases change implies revival of tradition. It is not easy to change the popular perceptions on the issues of caste, clan and social rank.

Desai (1969) linked the spread of modern means of communication, British administrative system and laws with a view to analyse the relationship between caste and class. According to him, the village undermined more and more the functional basis of caste. The transformation of self-contained rigid castes into modern mobile classes has taken place in a peculiar manner. This development has created the peculiar social structure in modern India.

Reconstructing the past geographies is necessary in order to develop an understanding of why the people behaved the way they have been behaving. The roots of most of our problems lie in the past. The way cultural differences have emerged are the necessary foundations on which a super structure can be built.[13]

Despite the process of urbanisation and the growth of infrastructure, the traditional social order based on caste hierarchy is still intact. In fact, caste hierarchy is very often exploited for electoral gains. Indeed, it is taken for granted that the caste-like formations, irrespective of religion, conform to a rural class system. It is interesting to note that the basic questions about caste, social hierarchy and rural development have never been addressed properly by the ruling elites of one hue or the other.[14]

What is underdevelopment? Underdevelopment is a direct result of the hierarchised social structure. Perhaps it is the product of on-going administrative laxity due to deliberate neglect. The politico-administrative set-up has often played a critically important role in keeping these identities intact. In fact, access to land is the basic issue for the agricultural communities. It is determined by the hereditary land rights defined by the custom/tradition. In rural India, for example, caste system places everybody in its own respective place. The evolving social system has assigned a specific social space for everybody. The over-arching traditional hegemony of the caste elites on land resources goes unchecked and unchallenged. The party politics during the last fifty years has further reinforced the practice of discriminatory caste system. It has contributed to the glorification of the hierarchised social structure and has added a new meaning to it. Caste-related institutions, such as the political-party formations, and the caste councils have sharpened the cutting edge and have emerged as a phenomenon. In the rural areas of the subcontinent, from Hindu-Koh Himalayas to the tip of the peninsula and from Herat to the Naga Hills, the inevitable conclusion is that the social structure plays a critically important role in defining the access to land. There is a well-established tradition, which protects the interests of the landed peasantry.[15] There is a method and a political logic in retaining the tradition as such.

Beteille (1992) pointed out that the extremes of privilege and deprivation are so characteristic of the agrarian society of Asia. He further asserted that the Hindu caste system formed the basis of social division. In his view, caste is particularly important in the social system of agriculture. This is especially true in the context of the skewed distribution of land.[16]

In the rural areas of the subcontinent, as in other parts of the world, agricultural economy is by and large determined by the environmental and ecological factors. All this shows that the rural scene has not changed fast. The same caste-class structures define the magnetic field of the process of rural development. Whether one can avail, or not avail the infrastructure created for rapid agricultural development, such as the improved seeds, agro-machinery, chemical fertilisers and assured irrigation, depends on the ownership of land and the prevailing tenurial system. The caste ideology provides a framework for the transformation of the agricultural scene. It is the parameter of land ownership, which differentiates between the rural haves and the have-nots. There are two models (packages): first, an iron plough, a pair of bullocks, a camel or a buffalo and a sickle; and secondly, a tractor, high yielding varieties of seeds, chemical manures and assured irrigation. Which of the two models would be adopted is an open question. But this characteristic determines the future of the rural masses as well as the future of the subcontinent. However, those who do not own land, agricultural infrastructure, such as the one described above, is irrelevant. The rural poor will continue as landless labourers dependent on the landed peasantry. The other option for them is 'out-migration'.

The Dream of an All-India State

The British Empire in South Asia (1757-1947) extended from Gilgit-Hunza and Baltistan in the northwest to Burma (now Myanmar) in the northeast and from the Wakhan territory in north Afghanistan to Ceylon (now Sri Lanka) in the south. The vast empire (area: 5.14 million sq. km. and population: 1,367.1 million) was administered and controlled through a dualistic system of governance, *i.e.*, directly controlled territory consisting of the British Provinces and an indirectly controlled amalgam of princely states, numbering more than 562 at independence

(Sukhwal 1985). This was the only stage in history when the subcontinent was in the control of a single authority – the British government of India. The Indian state after independence was a much smaller and reduced version of the British Empire. It has been observed by historians that an all-India Empire is not likely to last more than a few generations. 'The great empires of India that arose remained intact for not more than a century and a half, as against the 400 years of the Roman Empire and the two millenniums of China' (Basham 1964: 4). There are, of course, dissimilarities in the life history of the empires in India as compared to Mesopotamia, Egypt or China. The general view is that India failed to establish a single superstructure to sustain an empire under the charge of a single centralised ruler. The reasons for this 'failure' were supposed to be rooted in the character of the Indian climate, difficulties in communication, lack of sense of history and the other worldly nature of the Indian character. None of these explanations can stand the test of reason. The reasons cited as problematic were not new to India. All other empires, such as the Chinese Empire, faced the same difficulties. In later times, the legend of Digvijayins as the rulers of the entire subcontinent from the Himalayas in the north to Sri Lanka in the south has always been highly popular. In fact, according to Basham, 'it became a conventional hyperbole of royal panegyrics to claim that a powerful king had actually carried his arms to the very bounds of the Bharatavarsha' (Box 1.1). Whatever may be the reason, the dream of an all-India united state could not be fulfilled (Basham 1964: 6-7).

Box 1.1: Inspiration of the *Digvijaya*

....Thus throughout the history of Hindu India, though the land was broken into many kingdoms, the inspiration of the *digvijaya* was always present, and the ideal of a single hand controlling the whole of the subcontinent was never forgotten.... I believe that the main reason for the failure of Hindu and Muslim alike was that no Indian ruler or dynasty was capable of providing administrative machinery adequate to control a really large state (Basham 1964: 6-7).

Changing Political Organisation

British India consisted of directly ruled provinces as well as indirectly ruled princely states. The princely states enjoyed an 'independent' status in matters of governance (Joshi 1976: 4-5). They were free to collect land revenue (*lagaan/malguzari*), have their own currency, postal and revenue stamps, administrative machinery, police and the railways. The issues, such as succession and external relations with other states and with the British were managed and coordinated by the British agents. These agents were appointed by the Raj authorities to oversee law and order problems, transfer of power and other succession matters. The princes were bound by treaties to help the British in internal and external wars, including armed struggles in various parts of the world (Sukhwal 1985). The princes constituted a captive pro-British lobby, which helped the British crush the pro-independence movement led by the Indian National Congress Party. The princely houses enjoyed freedom to build palaces and forts. They kept an unnumbered cache of pedigree horses and dogs and a large number of wives and concubines. The British found these princely houses as cultural niches, and dependable loyalists. They enjoyed their hospitality and helped them retain in their royal ways of living. They also appreciated their patronage of arts and crafts, particularly preservation of classical music, establishing museums in capital cities of the princely states displaying the history of the lineage and characteristics of arts and crafts. They also invited musicians to play in their *darbars*. It was due to this patronage that numerous schools of Indian classical music (*gharanas*) survived and developed under the Raj. The classical Hindustani music suffered a major setback when the states crumbled after independence in 1947. The states were given three options: to merge either with India or Pakistan[17] or to remain free. The Raj also helped the princes modernise their cultural tradition, folk arts, and folk dances. The give-and-take between the British and the princely states resulted in the emergence of a new form of Indo-British culture. The British in the administrative services adopted the finer elements of the Indian culture and were greatly influenced by the life-style of the Indian aristocracy. Cocktail parties, followed by dinners comprising several courses, such as *biryani*, *murgh musallam*, and *kulfi*, to mention only a few, became a standard practice. Likewise, they also adopted indigenous methods of cooling the rooms during the summers using *khas tatti* as a common cooling device. These traits became symbols of a new cultural synthesis. Occasionally, the British also participated in the celebrations of Holi, Diwali and Dusshera.[18]

With independence in 1947 the princely states became the centres of diffusion of India's cultural tradition, which was handed over to the new generation of rulers who came to power in different parts of the subcontinent. The new government created centres/institutions for the preservation and promotion of elements of traditional culture, folk arts and classical music, such as Lalit Kala Academy, Sangeet Bharti, Sangeet Natak Academy, Indira Gandhi National Centre of Arts, All-India Fine Arts and Crafts Society, Shri Ram Bharti Kala Kendra, Triveni Kala Sangam and Yamini School of Dance, etc. Some of these institutions were also established by the business houses. The All-India Radio also provided patronage to the singers and artistes specialising in different styles of dance and music. Special programmes are broadcast and telecast on the All-India Radio and Door Darshan.

Evolution of Territory

Today the subcontinent is divided into several countries, viz. Afghanistan and Pakistan on the northwest, Bangladesh on the east, Nepal and Bhutan on the northeast, Sri Lanka in the south and the Maldives on the southwest. India, of course, occupies the centre stage and has been a centre of culture and civilisation. However, India also has played a significant role in the diffusion of the elements, which are typically Indian and define the personality of the subcontinent. The region has passed through several phases of a tortuous political history. Notable among these phases were the struggle for the independence of India and Pakistan, which came along with the partition of the country and the holocaust that preceded/followed. Certain parts of the subcontinent, such as Sri Lanka, Nepal and Bhutan emerged as sovereign states. Burma (now Myanmar) was separated from the British Empire in accordance with the provisions of the India Act of 1935. India, Pakistan and Bangladesh passed through several phases of recent political history. Things changed drastically after independence in 1947 with the integration of princely states through merger with either India or Pakistan (Menon 1956); later reorganisation of the states on the linguistic principle changed the political map of India. Today the position is that a substantive part of the territory of Jammu and Kashmir is in the possession of Pakistan. The Maharaja of Jammu and Kashmir agreed to merge with India under pressure from the Pakistani tribesmen who entered the valley of Kashmir. This generation of the people of the subcontinent has inherited a truncated Jammu and Kashmir state whose parts are now occupied by Pakistan and China. India and China also have different perceptions about the borders of Arunachal Pradesh. In recent years the Chinese seem to have abandoned their traditional policy on the regularisation of the disputed territory. It is alleged that Chinese do not accept the McMahon Line as a correct position on the ground. Today there are differences between India and China on the geographical position of the two lines of armistice, one in Jammu and Kashmir and the other in Arunachal Pradesh. Similarly, there are differences between India and Bangladesh on the question of common borders and on the question of the Bay of Bengal Islands off the Bangladesh coast or on sharing the Ganga waters. Claims and counter-claims have made the resolution of the problems difficult. It has been suggested that a perspective on the history of the subcontinent is necessary in order to understand the vexed problems of the South Asian states. Moreover, the political upheavals — four wars between India and Pakistan and one war between India and China — have left scars on the face of the land. The return firing across the borders of these nations has become a daily affair. Negotiations between India and Pakistan have been an on-going process. But firing across the Indo-Pakistan border continues unabated. The difficult terrain in the state of Jammu and Kashmir makes the defence of the country a highly complex exercise. The difficult terrain also offers opportunities to the militants whose infiltration has caused an erosion of the confidence between the two nations, viz. India and Pakistan. Likewise, the eastern borders of India with Bangladesh have not been free from cross-border skirmishes.

The problem of political adjustment within the subcontinent and between the nations has several dimensions. Geographically, the northwestern and the northeastern parts of the subcontinent have a backlog to be cleared as early as possible. The two lines show that there is a sufficient scope of conflict along these lines. It is expected that the problem would be sorted out on the basis of mutually agreed solutions. It would be in the interest of the people of the subcontinent if these issues were settled amicably and through peaceful negotiations without any conditionalities. The present tactics can no longer be accepted as a policy plank. As pointed out earlier passing the buck to the coming generations may be risky.

Contrary to the common belief held by the political and historical geographers, the subcontinent is not an isolated territory. In fact, it has always been a theatre of action and reaction among the peoples for centuries altogether. As we know, Iranians (Parthians) and Greco-Bactrians were controlling the entire northwestern region from Afghanistan to Jumna (now Yamuna) and from Kashmir to Kathiawar in Gujarat. The on-going process of cultural intermingling and the adoption of the British administrative system, a civil law as well as the conventions of parliamentary democracy have brought the South Asian nations to a common platform.[19] However, the parliamentary democracy has not always been a successful experiment. India, the largest country in the region, has strived hard to keep the principles of democracy intact. This may not be true for the other nations in the region. The people and the states had their own specific reasons to exercise their options. Barring a short period of twenty-five months (June 1975 – July 1977), when all democratic institutions were suspended and the print media were censored ruthlessly, a democratically elected Prime Minister of India — Mrs. Indira Gandhi – was forced to impose emergency in June 1975. However, the nation came out of the nightmare fairly successfully. On the other hand, democracy has not been a successful experiment in Pakistan. The military junta has acquired power through recurrent *coup d'états* time and again. Military leaders from Major General Ayub Khan to General Parvez Musharraf have ruled the country with a larger measure of success than the democratically elected prime ministers.[20] In Pakistan there is a thin edge between democracy and authoritarianism. Some of the South Asian nations have always been swearing by democracy but not adhering to the basic tenets of a democratic polity. In the sixty years of the history of Pakistan since its inception in 1947, military leaders have assumed absolute power. They have always taken shelter under the Islamic doctrine of justice and republican conventions. They might have been successful in keeping the masses in good humour in the name of Islam, but only superficially. It has also kept the Islamic doctrines going smoothly in order to satisfy the people at home as well as the other Muslim countries abroad. These nations sympathise with Pakistan because of its alleged adherence to the Islamic law, whatever it implies in a modern democratic polity[21] and the system of governance based on it.

There are reasons why a reconstruction of the past geographies is necessary to develop an understanding of why the people behave the way they are behaving. The roots of most of our problems lie in the past. The policies for abiding peace have laid down the foundations on which the superstructure of peace can be built.

A.R. Desai (1969) and other sociologists, such as Joshi (1976); political scientists, such as Khan (1992, 1997) have reflected on the issue of identity based on caste. While it has been observed that there is a difference between caste and class, Desai argued 'that the process of development has created a peculiar social structure...with the result that in the existing Indian society, class struggles have been often assuming the form of caste struggles'. The caste identity is visible as compared to the class, by real or imagined. Provisions were made in the federal Constitution of India to safeguard the interests of the lagging sections of the society who largely conform to the lower rungs of the caste system.

It may be reiterated that the social structure plays a critical role in defining the personality of the nation states of South Asia. People behave in a manner, which is typically Indian. The state, particularly in India is always found engaged in a balancing act of sustaining the diversity in order to ensure unity. The social structure evolves a particular polity best suited to its needs. It is a little wonder as to how the state in South Asia functions to satisfy the diverse elements of society. There are vast differences in the ethno-lingual characteristics of the people from Pushtuns in northwestern parts of the subcontinent to Nagas in the northeastern parts of India. In fact, there are little and great culture-worlds which have survived the rise and fall of empires. The governments have to learn the art of walking on a tight rope in order to ensure their continuity and to defend the interests of all the groups making up the society. Understandably, it is not easy to satisfy all the sections of the society either in Pakistan or Bangladesh and Sri Lanka. In India also, the act of sustaining the state is too complex. Regional ethno-lingual groups, sometimes in hundreds, struggle to retain their identity. The job of the central government is to keep all elements together and induce them to play their role in a constructive manner, a manner so that the viability of the state is not eroded by the political exigencies. We have seen a scramble for power by groups, which have political aspirations. Our experience since independence shows how the tussle between regional/parochial identities has been managed even at the cost of the national level solidarity. During the period since independence the national parties have lost ground to the regional parties. Each state of the Indian union is believed to be a cultural unity. The Congress Party has gradually lost its hold on the masses in several parts of the country. The glorious period of the Congress Party's rule eventually eroded with the growth of regional political parties based on lower level differences in aspirations and claims to territory. States have been reorganised time and again to satisfy the political aspirations of the local/regional communities. This has happened due to the concentration of caste, ethno-lingual homogeneity and class structure. These conflicting political aspirations have brought the Indian democracy to the stage of turmoil with the result that only coalition governments can function. It has been noted that twenty-three or more parties came together to form the coalition government in New Delhi, such as the NDA (National Democratic Alliance). The regional leadership is vocal and vociferous about the local/regional issues. However, this model is not universal. For example, Pakistan and Bangladesh are far more politically stable because of homogeneity in cultural identity. But democracy has not been functioning satisfactorily. Both the countries have experienced a seesaw battle between the civilians and the armed forces. More often than not the military regimes have been more successful than the civilian democracies. Pakistan may have far more complicated problems in managing the political aspirations of the regional ethnic groups. There is little progress on the issue of keeping together the groups, such as the Baluchis, Pashtuns, Sindhis and the Punjabis. The state in Pakistan has gone in the hands of armed forces because the people belong to four or five nationalities (read regional ethnic groups) and they can be controlled only by force. The Punjabis in Pakistan have been successful in having a larger share of the national apple-pie. Baluchis, Seraikis and the tribal *jirgas* have always complained about the Punjabi hegemony on the natural resources of the state. Bangladesh, on the other hand, enjoys a unity because everybody comes from a Bengali background. The presence of Biharis is not welcome. However, military rule has brought the Generals to rule through the *coup d'états*. The Generals have also been assassinated and often replaced by their widows, or other close relations, through democratically contested elections. They have won the elections riding on the sympathy wave for them. Unfortunately the rule of the Begums is likely to be short-lived. The relationship between social structure and polity should be studied more carefully. Afghanistan, Nepal and Bhutan have experienced the monarchies, a form of dynastic rule. It is in this context that the present political situation should be viewed.

END NOTES AND REFERENCES

1. An English poet of the nineteenth century expressed his wonder when he wrote these lines:

   "Where are you Time and Space,
   The World is a Little Place."

2. The policy of opening up the country to foreign investors and governments responding to the process of globalisation has brought the Multi National Companies in the free market economies of the nations of the subcontinent. This has brought about a total revolution in the arena of economic development. Disinvestment policies further indicate withdrawal of the governmental control and regulatory systems. See Mahbub-ul-Haq, *Human Development in South Asia – Globalisation and Human Development*, Oxford, 2002.
3. It appears that the Hartshornian logic of spatial variation and areal differentiation is now irrelevant, due to the impact of new methodologies of research available now. Geographers, appear to be in transit from simple to the complex. See Hartshorne, *Nature of Geography*, 1939.
4. European Union (EU) is a cartel of European nations. It espouses the cause of concerns and collective action to promote the common good. It is a collective effort to meet the challenges of the new world economic order, and avoid unnecessary competition between the nations of the region.
5. The people in the region are reeling under the fury of discriminatory prejudice and poverty. Sri Lanka and Pakistan may have a slight edge on India and Bangladesh in terms of social and human development. In India, the problem of squalor, health and habitation are overwhelming and the political regimes in power have little to offer to the marginalised masses.
6. The philosophy of humanism is based on the firm belief that common human needs are to be addressed in a way so as to promote the common good 'seeking solely rational ways of solving human problems and concerned with the mankind as responsible and progressive intellectual beings'. *Concise Oxford Dictionary*, 1990: 575.

7. *Dalit* is a Hindi word, which means oppressed or persecuted. The low castes in India are generally described as *dalits*. The word has acquired significance in the recent past, particularly in social science research.
8. The term scheduled castes applies to the caste groups mentioned in a certain schedule of the Indian Constitution, 'those beyond the pale of *Varna* system', Srinivas in K.S. Singh, *People of India, Anthropological Survey of India,* Vol. I, 1992: 5.
9. 'Ours is a country which is very rich in our diversity, and there is often talk of tolerance of diversity.... I think we should appreciate that our glory lies in our diversity', Srinivas, in K.S. Singh, *People of India, Anthropological Survey of India,* Vol. I, 1992: 4.
10. *Ibid*: 4.
11. International economic order generally refers to the terms and conditions which the super powers impose on the flow of goods and commodities between the countries and the limitations of the tariff applicable to the little economies and small countries.
12. In global economic reports, very little or no reference is made to the non-performance of governments, particularly in the developing countries. The political leadership and the captains of business and industry of these countries are directly responsible for the mess. Electoral rhetoric is the rule of the game. Development funds are freely diverted from one head to the other. It is alleged that political leaders make promises in public meetings and allocate funds for development projects or for relief work without any reference to either the National Planning Commission or the State Planning Boards.
13. Increasing spread of the modern means of communication, the introduction of the British system of administration and laws, and the growth of modern, capitalist, competitive economy, which shattered the subsistence economy.... The village undermined more and more the functional basis of caste. However, the transformation of self-contained rigid castes into modern mobile classes has taken place in a peculiar manner. Certain castes have been monopolising the position of the privileged upper classes of the modern society.... This development has created the peculiar social structure in modern India, with the result that within the existing Indian society, class struggles have been often assuming the form of caste struggles (Desai 1969).
14. The governments in the South Asian countries have always been concerned about the problems of rural development. It is evident that the progress in this sector continued to be a major policy plank of all state and central governments since independence. 'Caste has been particularly important in the social system centring around agricultural production, especially the control and use of land', Ahmad, 1980, Proceedings of the Second International Symposium on Asian Studies, Vol. II, pp. 905-922, Asian Research Service, Hong Kong.
15. Communities involved in agriculture in the subcontinent consist of mostly landed. However, there are landless who constitute the landless workers. The landed control the traditional rights to own land and cultivate on it. Despite the land reforms a general impression is that the landless have not been the beneficiaries of these reforms.
16. '...India represents more sharply than any other country the extremes of privilege and deprivation, which are so characteristic of the agrarian societies of Asia. The Hindu caste system, which formed the basis of social divisions in traditional India, may be viewed as the proto-type of a hierarchical system. Caste is particularly important in the social system centring around agricultural production, especially the control and use of land' (Beteille 1992).

17. While India decided to send armed force to subdue the Nizam of Hyderabad to accede to India as well as to crush the *Razakar* movement, the Maharaja of Kashmir was left untouched. During these critical seventy-two days tribesmen raided the Kashmir valley and captured the territory of the Dogra Maharaja lying beyond the North Kashmir Range. See also B.L. Sukhwal, *Modern Political Geography of India*, 1985: 23.
18. The Maharajas of Mysore had a long tradition of celebrating the Dusshera festival. While the state of Mysore was merged with India, Dusshera continued to be celebrated with the same pomp and show.
19. It is interesting to note that while India has remained committed to the parliamentary democratic system as the state policy, this cannot be said for Pakistan and Bangladesh. There were several phases when the military leaders captured power to rule these countries.
20. There is a relationship between democracy and dictatorship in the two countries of Pakistan and Bangladesh. The military regimes have introduced a form of democracy, which is not free but guided. For example, General Ayub Khan of Pakistan introduced a system of 'guided democracy'.
21. Societies adopting the Islamic Law as a state policy find themselves tied down to a certain interpretation of the Islamic doctrine. But the issue seems to be disputable.

2

The Subcontinent: Nations, Nationalities and Nationalism

The South Asian subcontinent is also described as the Indian subcontinent. In this chapter an attempt has been made to introduce the region (Maps 2.1 and 2.2). A brief country profile has been presented here to understand the regional setting, nations and nationalities. Geographical factors, as far as we know, have played a role in defining the personality of the South Asian region. The inter-country relations have been described in order to realize the nature of the diversity in unity. People in different countries of the subcontinent have a shared history. However, the international relations have not been free from problems. The differential levels of economic and human development have often resulted in divergent perceptions. Even though there are sharp differences in personal names, dialects, dress ways and culinary habits. The cultural characteristics and the ethnic composition of population show an interesting mix. The South Asian nations have been cast into a mould, which is characteristically subcontinental and unique in its basic traits.

In this chapter we have tried to turn a few pages in the recent history of South Asian nations from Afghanistan to Maldives. The purpose is to understand the past in order to ensure an amicable present and a better future, free from civil strife and peaceful international relations. An attempt has been made here to capture the elements, which describe the geographical personality of the subcontinent and acts as the unifying principle. There are diversities in the physical make up of land (structure, relief and drainage), the evolution of territory, history of peopling and the contemporary political and international relations between and among the nations of South Asia. The history of South Asia is full of references to the process of unification and subdivision and re-unification. Empires have been rising and falling leaving behind the primeval states/republics, or kingdoms. People have forgotten their past and the story of rise and fall is not a part of their memory. People often forget their past history but the folk tales of this rise and fall have always been preserved in their subconscious mind. One of the major traits of the people of the subcontinent is glorification of the past. Much of this folklore has entered into the minds of the authors of the bardic literature.

Afghanistan

Afghanistan is often described as the land of hardy, freedom-loving and trigger-happy Pathans, also known as Pakhtuns or Pushtuns.[1] Afghanistan has a congenial climate, which offers scope for the cultivation of a variety of fruit crops, such as grapes, almonds, walnut and pine nut (*chilghoza*). The country is also known for its woollen shawls, karakul pelts and rugs. For centuries Afghanistan has served as a gateway to India for travellers from Central and West Asian countries. It is here that some of the major countries of Asia meet – India, Pakistan[2], Iran, Turkmenistan, Uzbekistan, Tajikistan, Kirghistan and China. On the north lie Xinjiang and Tibet, autonomous provinces of China. The corridor role of Afghanistan is not only the most significant feature despite its land-locked status; it has provided the space for racial and ethno-lingual intermixing. Historians, such as Toynbee (1961) have recognised Afghanistan as one of the 'most important roundabouts of the old world'. The Kabul valley route has always been a conduit through which the traffic of people intending to reach India has passed. It has enabled the people of West Asia and Central Asia, China and the regions lying in the east and the west of the Pamir promontry[3] to intermingle here. It is in this role that Afghanistan has served as a major meeting ground throughout the history of the old world.

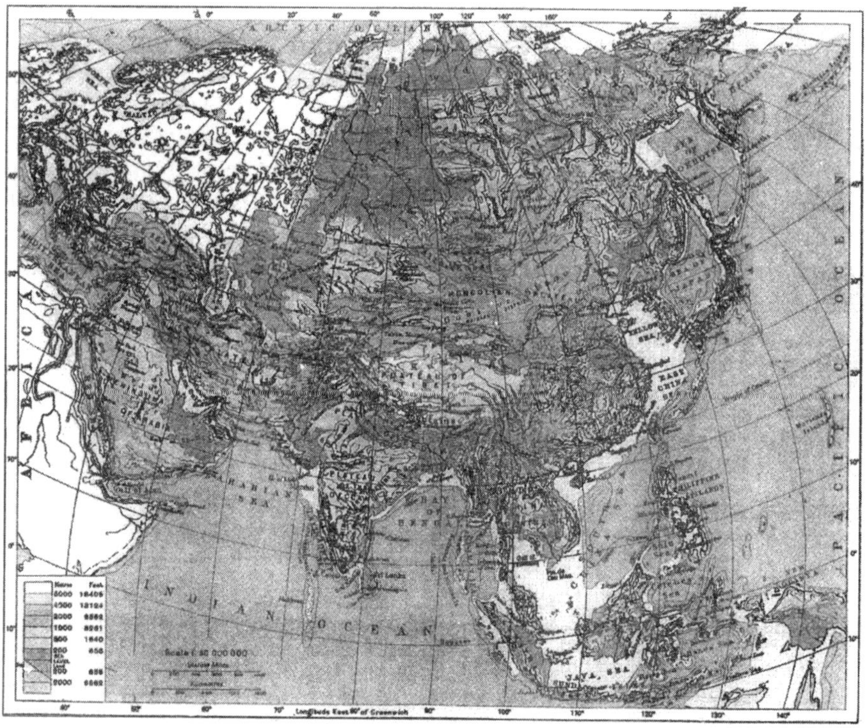

Map 2.1: The Continent of Asia

Roughly elliptical in shape, the country is characterised by uneven surface features, such as high mountains and intermontane valleys (Map 2.3). The terrain with an average altitude of 4,270 m. (14,000 feet) enjoys a moderate temperate type of climate. The most significant relief feature is the Hindu Koh.[4] The mountains dominate the landscape of the subcontinent and define its northern limits. They have always determined space relations with the rest of the world. The Hindu Koh extends for a distance of 800 km. from the Pamir plateau to the Iranian borders in a northeast-southwest direction. The mountain complex includes besides Hindu Koh, Karakoram and the Himalayan systems. These mountains define the northern and northwestern limits of the subcontinent. On the north, the Hindu Koh range gives way to Karakoram and the Himalayan mountains. Some of the peaks are as high as 7,620 m (25,000 feet). Most of the area is mountainous. The Hindu Koh Range is the source of the Kabul river and some of its tributary streams. The Kabul river carries the snowmelt from the highlands and empties itself into the Indus river near Attock. The Amu Darya (Oxus) forms the boundary between Afghanistan and Uzbekistan.

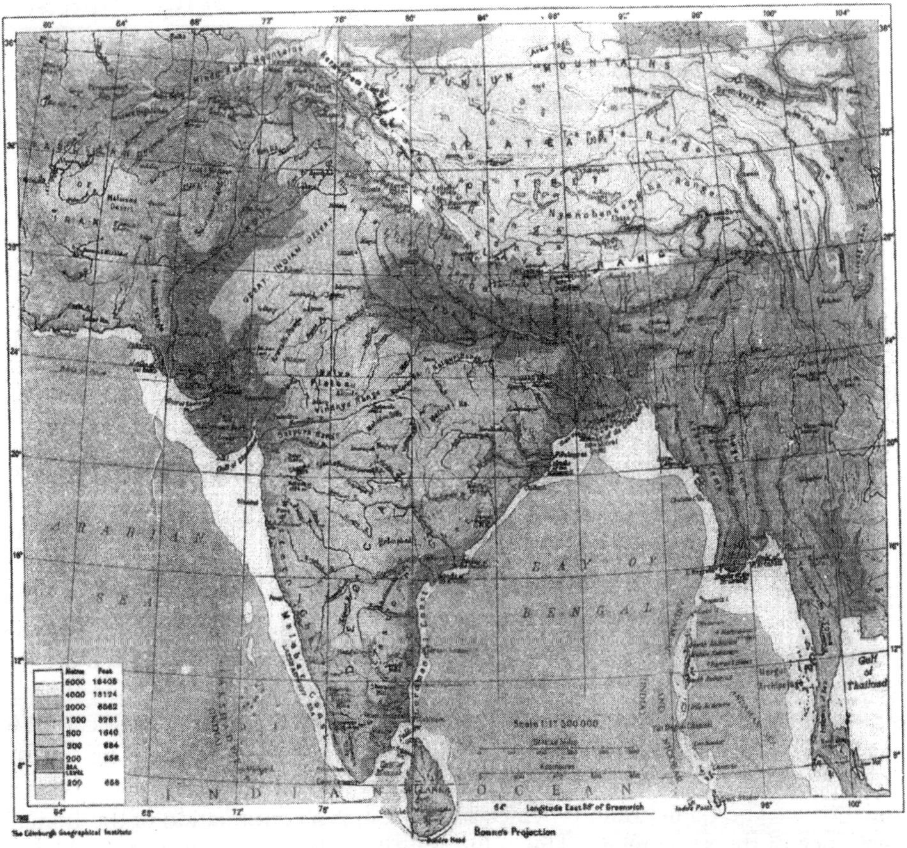

Map 2.2: The South Asian Subcontinent

Source: http://atlas.mapquest.com/atlas/

Map 2.3: Map of Afghanistan

The famous city of Mazar-e-Sharif on the Afghan side and the city of Termez across the river in Uzbekistan are the outposts of the two countries. Movement across the Hindu Koh Range has always been difficult but not impossible. It is due to the high altitude of the passes, such as Shibar pass (2,987 m. or 10,000 feet), that the traffic of humans and animals has always been thin. The Shibar pass connects Kabul with the northern outlying regions of Afghanistan. The terrain in this sector consists of lowlands, much valued for the agricultural potential. The famous Khyber Pass lying in the southern extension of the Hindu Koh Range, locally known as Safed Koh, has served as the main gateway to the northern and western parts of the subcontinent.[5] Similarly, the Bolan Pass situated in the Kirthar Range offers access to the neighbouring parts of Baluchistan in Pakistan. The southern and southwestern parts of Afghanistan are too dry and inhospitable. This part is dotted with lakes and swamps, a typical example of inland drainage. Helmand is by far the longest river, which empties itself in the lakes and swamps of the dry region, also known as *registan*.[6] Afghanistan has passed through several stages of civil strife and its recent political history has disturbed the peace and tranquillity of the Afghan people. The problem arose when the erstwhile Soviet Union allowed to establish a communist regime converting the monarchy into a political party system. Much of the Afghan blood fled during a long drawn struggle. The Taliban regime, which was deeply religious, imposed its policies of Islamic revival and a civil strife. The Baluchistan province of Pakistan is a continuation of the Iranian province of the same name. The cultural history of Afghanistan is a product of interaction between different tribal groups living in this area.

Pakistan

Pakistan came into being as a result of the partition of the subcontinent in August 1947.[7] The two states that were born in the midnight of 14-15 August 1947 were identical twins, equal in political status, unequal in size and weight (Map 2.4). They carried with them similar congenital problems. While Pakistan was anaemic at birth, India was found to be suffering from chronic underdevelopment and regional disparities in economic growth and human development.

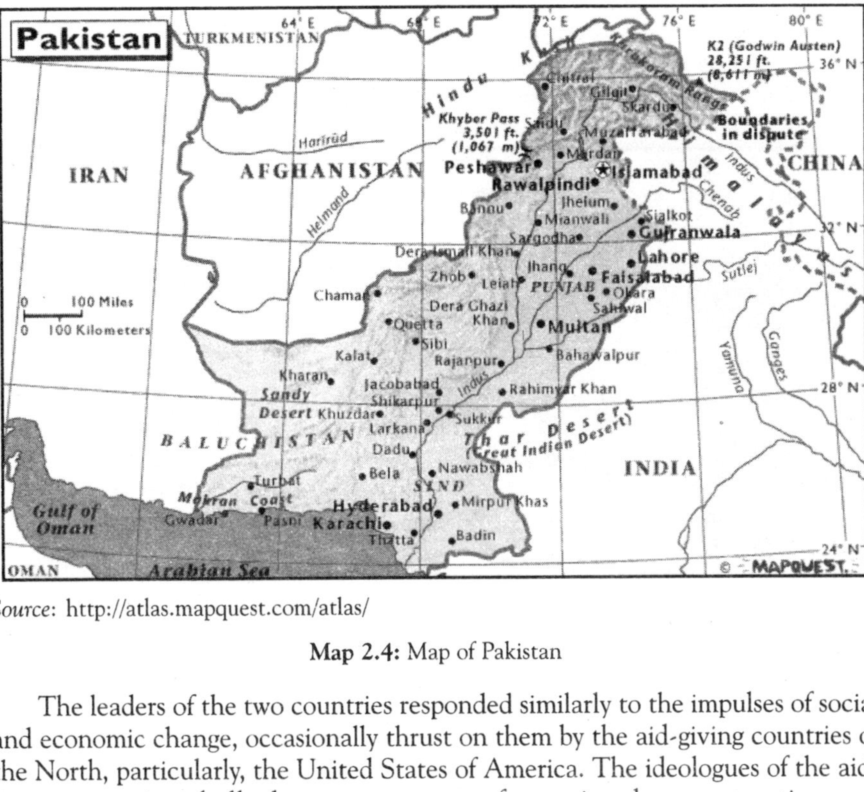

Source: http://atlas.mapquest.com/atlas/

Map 2.4: Map of Pakistan

The leaders of the two countries responded similarly to the impulses of social and economic change, occasionally thrust on them by the aid-giving countries of the North, particularly, the United States of America. The ideologues of the aid-giving agencies chalked out programmes for national reconstruction and comprehensive development. Aid giving agencies came in bundles making every Indian and every Pakistani national indebted to the aid-giving agencies of the Northern world (read developed world).

While Pakistan was a dreamland of the leaders of a section of the Muslim community of the subcontinent in the thirties and the early forties of the twentieth century, the communal climate preceding the independence and partition of India converted it into a ground reality. In fact, the communally charged atmosphere was one of the most important causes of partition. The idea of Pakistan did not mean that all the Muslims on this side of the newly carved-out borders would migrate to Pakistan, East or West and that all the Hindus living in northwestern India before the country was partitioned, or those in East Bengal would migrate to

India. The circumstances were, however, such that millions of people crossed the borders from either side. Some did so for fear of violence, which was going on intermittently all around and some in apprehension that the things may worsen and more violence may erupt anytime anywhere. History proved that their fears were not unfounded (Box 2.1).

Box 2.1: On Partition

This might be taken as an exemplification of the working of some persistent and mysterious law of Indian history, which effectively prevents the permanent establishment of a single government over the whole of the subcontinent. But, in fact, the partition of 1947 was due to forces, which had not been significantly operative in earlier times. In the previous cases of the break-up of Indian empires similar, though far from identical, causes – for instance the absence of a sound administrative tradition and difficulty of communications – seem to have been at work. But in 1947, a tradition of efficient administration was established and communications were adequate. Another factor had arisen, which completely altered the situation. The growth and fruition of the idea of Pakistan was something quite new in India, with no precedents in her earlier history, and the partition was altogether different from anything, which had happened before (Basham 1964: 13).

Suggestions have been made in political and quasi-political circles in India that the task of partition is not yet complete and that it would not be considered as having been accomplished until the last Muslim soul leaves the country for a destination either in Pakistan or elsewhere. Perhaps this opinion did not find favour from either Mahatma Gandhi or eminent Congress party leaders such as Nehru, Sardar Patel and Maulana Azad. The situation is viewed as explosive when Muslim Bangladeshis are located by the police in the Indian states of West Bengal (Kamaluddin 1985: 221-36), particularly in the border districts or in the metropolitan cities, such as Calcutta (now Kolkata), Bombay (now Mumbai) and Delhi or in other parts of India, such as Assam, Meghalaya, Tripura, Bihar, Uttar Pradesh and Chhattisgarh states. It is feared that the deportation of such elements is on the cards. The fear psychosis of the pre-partition period has been reborn and is very much in place. This is because of the fear of Muslims becoming a majority due to rapid population growth haunting the mind. It is further assumed that the day is not far away when the Muslims would outnumber the Hindus.[8] However, it appears to be an exaggeration of the ground truths. The process of purification, it is suggested, should be expedited to cope with the over-arching problem of the growth of Muslim population. Demography has its own blues.

The idea of Pakistan also had an implicit motive of purification. The word Pakistan is a compound noun made up of two words – *Pak*, meaning pure and unadulterated, and *istan*, a Persian word, meaning place, abode; the Hindi equivalent is *sthan*, place or *jagah* (location or territory). There are millions of non-Muslims in Pakistan and in Bangladesh. Keeping in view the fact that the world is becoming a unified village, in the wake of globalisation and *perestroika*, the motive of cleansing appears anachronistic, and an almost impossible task. The northern and the western parts of the subcontinent, including Afghanistan, and parts of Pakistan, such as Baluchistan, Sind, and Punjab have been more open to external influences. Vast chunks of the territory, such as Kashmir, Punjab, Sind and parts of Gujarat, Kachchh and Kathiawar, were in the control of alien rulers—Greeks, Greco-Bactrians, Iranians (Achaemenid), Arabs, Turks, Uzbeks, White Huns, Indo-Scythians, Sakas, Kushanas, and the like. The territory of Pakistan, when it emerged as a separate country in 1947, was genetically linked to the occupation of this part of the subcontinent by external powers. The Greeks and Greco-Bactrians ruled for no less than two hundred years. Earlier the Iranians Achaemenid Empire had ruled over this region for about two hundred years. The demand for Pakistan was the culmination of a long history of separatism and communalism turned into hatred. The nuclei of political power based in Patliputra, Kanyakubja or Agra, Delhi or Hastinapur were not always in a position to control the northwestern parts of the subcontinent for more than brief periods of a few centuries. A careful study of the history of the northwestern part of the subcontinent suggests that it was easy for the people in these regions to respond to the centrifugal forces. The territory was too vast to be controlled from a single centre of power. It may be noted that the on-going conflict between India and Pakistan is a kind of a rebirth of the spirit of distrust and dejection, expressed in the never-ending internecine war between and among the petty kingdoms in the ancient and medieval periods of our history (Basham 1964: 14). The twin states of India and Pakistan behaved similarly. If one test-fired inter ballistic missiles (IBM), the other was not lagging behind. This has resulted in a never-ending race for arms, including the nukes. Today, the ground reality is that both the countries are prepared for any eventuality in the arena of war.

The emergence of Pakistan as a sovereign state in 1947 out of the British Empire in South Asia was in a way, a little wonder of the world. It had two wings—East and West, separated by 1,600 km. of Indian territory. There was no opportunity available to the people of the two wings of Pakistan to understand each other's social, cultural and ethnic traits. Nor were they prepared for an understanding of each other's social history. Bengali culture was miles away from the Punjabi, Sindhi, Baluchi and the Pushtun cultures. There was a sudden realisation that all of them were Muslim and that could be the basis of a separate state. The same authority of the Raj ruled the two regions but that was not a sufficient condition for unity.[9] The differences were more glaring than initially envisioned by the theorists of the idea of Pakistan. The people of the rice-eating East and the wheat-eating West were oblivious of the fact that the language and literature divided them more significantly than religion. Both Tagore and Nazar-ul-Islam, were, by and large, alien to the people of Pakistan in the West Wing, just as Iqbal and Hafeez Jalandhari were not familiar names in the memory of the people in the East Wing. They were diverse in looks, habits, culinary tradition and food preferences. They were also different in dress ways and aesthetic values. The galaxy of writers and poets who influenced popular thinking were different from each other.

The East-West relations deteriorated with time. In fact, not much time was wasted before the East wing broke away from the West. This was a test of the idea of Pakistan. Islamic fraternity and solidarity is not a game of numbers. This happened just after 24 years of living together. This new Bengali fortress was christened as Bangladesh.[10] The newly created state of Bangladesh took away about 16 per cent of the Pakistani area and 48 per cent of her population. The Bengali *Bhadralog* in the Indian state of West Bengal were amused by this development, but they were not enthusiastic about it. They were, however, nostalgic about their roots in the land now renamed as Bangladesh. They were jubilant at the fall of the Pakistani establishment in their immediate neighbourhood.[11]

Geographers generally recognize three major physiographic divisions of Pakistan—the Indus Plain, the Baluchistan Highlands and the submontane regions in North Western Frontier Province (NWFP). The Indus Plain (80 km. wide and 320 km. long) is further divisible in the Punjab Plain and the Sind Plain at the meso level. One observes a coastal plain, Akheron basin to the west of Baluchistan Highlands and the Thar Desert to the east.[12] Baluchistan Highlands consist of several mountain ranges, such as Toba Kakar Range, Sulaiman Range and the Kirthar Range. The renowned Khybar Pass, which was crossed time and again to reach India by travellers from central Asia and west Asia, is situated in the Safed Koh Range, a segment of the Hindu Koh Range. Similarly, the Bolan Pass east of Quetta provides a connecting route to the lower Indus basin. Pakistan consists of several provinces—Sind, Baluchistan, Punjab and the North Western Frontier Province (NWFP). Besides, there are tribal areas, which are administered by the federal government. The tribal areas are mostly situated in the northwestern part of Pakistan along the Pakistan-Afghanistan border. The federal government of Pakistan has not found it easy to control the tribal areas and their *Jehadi* movements based on the Pakistani soil.

Bangladesh

The state of Bangladesh is the latest addition to the political map of the subcontinent (Map 2.5). The most striking feature of Bangladesh is its physical environment. The economy and the society can be shaken at the slightest pretext. The country has a history of floods, tropical cyclonic storms and other natural hazards. These disasters have often led to the ouster of governments. Likewise, social upheavals are no less significant. First, the myth that commonality of religion can be the basis of state formation has exploded. East Bengal (East Pakistan) created on the basis of the so-called Muslim statistical dominance has lost its meaning. In fact, cultural factors, such as the Bengali language and literature define its identity better.

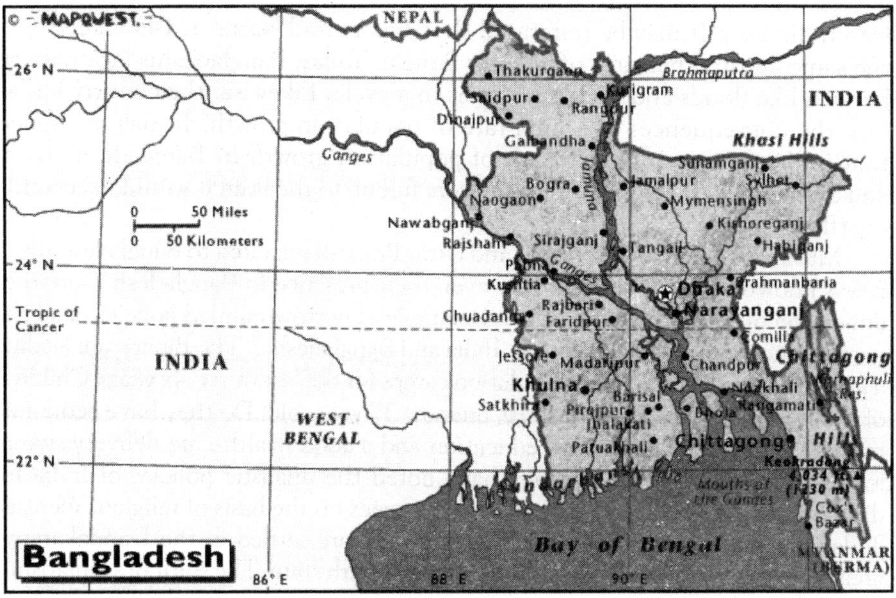

Source: http://atlas.mapquest.com/atlas/

Map 2.5: Map of Bangladesh

One may argue that the only difference between these two regions is religion. While West Bengal is a Hindu dominance area, Bangladesh is an area of Muslim concentration. By the same logic one may suggest that the Indian state of West Bengal should have been merged with East Bengal. This leads to the assumption that a sovereign state of Greater Bengal would have been ideal for the peace. This is an idea, which could not be sustained in the midst of prevailing atmosphere of confusion on the question of Bangladeshi identity. India played a role to further strengthen the state based on the religious identity. In our view, Bangladesh was a reincarnation of the idea of Pakistan. It obviously confronted with the Indian concept of secularism. In rejecting the logic of religion-based identity, and proving its inefficacy as a viable basis of state formation, India forced the Pakistani establishment to quit. Understandably, and as some critics pointed out, the change of regime did not uphold the basic ideal of secularism. In those momentous years, it is generally believed that the Indian leadership failed to realise that the religious basis of the newly created state of Bangladesh could be a source of tension. It could be a dangerous tool to play with. The uneasy history of inter-community relations in Bangladesh (Hindu Bengalis *vs*. Muslim Bengalis) proved this fact convincingly.

In order to highlight the fact, one may assume that the new state of Bangladesh was born in the womb of Pakistan. Out-migration of the poverty-stricken masses into India was a direct corollary of the religion-based politics. These questions would be asked and answered by the future historian. India helped the people of Bangladesh in 1971 but they did not imagine that the Indo-Bangladesh borders were soft and making infiltration of the Bangladeshis into India easy. It may be reiterated that Tagore and Nazar-ul-Islam belong to the same cultural heritage of Greater Bengal. Today, Bangladeshis face natural hazards like floods and cyclonic storms in a cycle. Likewise, the country has to face the consequences of a high rate of population growth. It may be argued that infiltration is a direct result of population growth in Bangladesh (Elahi and Sultana 1985: 15-35). If the cup were full up to the brim it would necessarily overflow.

Muslim population from Bihar and Uttar Pradesh migrated to Bangladesh when it was known as East Pakistan. However, their presence in Bangladesh is another dimension of the problem and a good example of narrow-minded policies of all the partners in this enterprise: Pakistan, India and Bangladesh.[13] The Biharis are leading a miserable life in the makeshift isolation camps for the last thirty-six years. Children of the Bihari families, born in 1970, are now 37 years old. Do they have economic means to survive, or seek good education and a good health-care delivery system for them? International observers have noted the dualistic policies of India for discrimination between the Bangladeshi refugees on the basis of religious identity. Today, the situation is that the Hindu refugees are settled in the Dandakaranya area in the then Madhya Pradesh, now Chhattisgarh state. The spirit of secularism[14] suffered a setback.

Nepal

The Raj authorities recognised the sovereign status of a number of kingdoms and chieftainships on the northern and the northeastern outskirts of their empire in India. The empire extended from borders of Afghanistan to Manipur and Tripura. These states were situated in the sub-Himalayan and the cis-Hindu-Koh regions and constituted a sort of buffer zone between the British and Tsarist Empires. The international boundary of Nepal extended from the *terai* in the south and the sub-Himalayan region in the north. It was in the *terai* region that the kingdom of Nepal faced India. Evidently, the competition between the Indian and the Nepalese peasantry intending to settle, or already settled in the *terai* region, was one of the reasons which contributed to an uneasy peace on the borders. Both the nations have complained of encroachment on either side[15] of the international boundary. The Himalayan kingdom is a hilly region and the scramble for level land in the *terai* region is understandable. Recent data show that migration from the hilly region to the *terai* region of Nepal has been a common phenomenon.

The Himalayan kingdom of Nepal is divided into: Great Himalayas, Middle Himalayas and the *terai* belt which runs parallel to the axis of the outer Himalayan range known as the Siwalik Range (Map 2.6). As noted earlier, the Great Himalayas are a formidable mountain range, with an average height of 4,570 m. (15,000 feet) and above. The Middle Himalayas, locally known as the Mahabharata Range in Nepal, is relatively not as high as the Great Himalayan Range, the average height being 2,500 m. (8,200 feet). The wet and once heavily forested *terai* region is situated in the southern part of Nepal bordering India. The relief features have

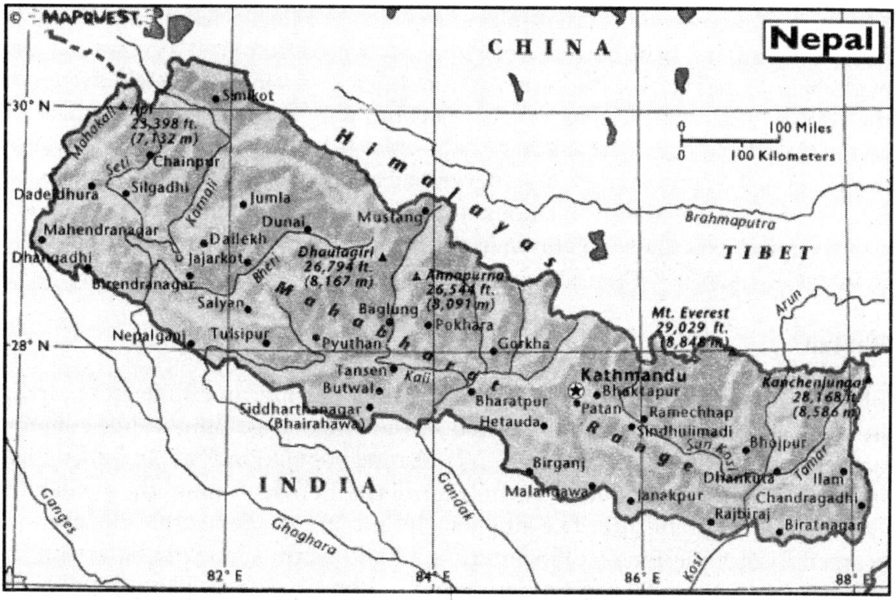

Source: http://atlas.mapquest.com/atlas/

Map 2.6: Map of Nepal

played a role in defining the zone of population concentration. These relief features have imposed limits on the human activity from lumbering and nomadic pastoralism in the north to agriculture in the *terai* region. Kathmandu valley is an attractive region because of its congenial climate. It excels both in terms of climate and agricultural land use. Nepal lies between the two great countries of India and China.

The Himalayan kingdom of Nepal is divided into: Great Himalayas, Middle Himalayas and the *terai* belt which runs parallel to the axis of the outer Himalayan range known as the Siwalik Range (Map 2.6). As noted earlier, the Great Himalayas are a formidable mountain range, with an average height of 4,570 m. (15,000 feet) and above. The Middle Himalayas, locally known as the Mahabharata Range in Nepal, is relatively not as high as the Great Himalayan Range, the average height being 2,500 m. (8,200 feet). The wet and once heavily forested *terai* region is situated in the southern part of Nepal bordering India. The relief features have played a role in defining the zone of population concentration. These relief features have imposed limits on the human activity from lumbering and nomadic pastoralism in the north to agriculture in the *terai* region. Kathmandu valley is an attractive region because of its congenial climate. It excels both in terms of climate and agricultural land use. Nepal lies between the two great countries of India and China.

It appears that the crux of the political relations lies in a balancing act of shrewd diplomacy, maintaining the policy of non-alignment with any of the two neighbours, who have a history of conflict between them. The policy of non-alignment between the two countries has helped the Nepalese nation immensely.

The geographical location of the kingdom has made it a fragile buffer state, receiving developmental aid from both the neighbours. The geographical location has also made Nepal a heaven for the fugitives from India, ever since the establishment of the British Empire in the South Asian subcontinent. The Nepalese also imported conflicting ideologies – parliamentary democracy from India and Maoism/socialism from China. The problems of underdevelopment and poverty of the Nepali people is due to the quasi-feudalistic relations of production. The Nepali youth are supposed to have convinced that communism is the only alternative for the liberation of the marginalised masses[16] (See also Appendix I).

Bhutan

Bhutan derived its name from the Bhotia or Bhote tribe, an ethnic group, living in India as well as in Bhutan. Nestled in the Himalayas, Bhutan has common borders with India and China (Map 2.7). A total population of about 0.67 million is distributed unevenly due to the hilly terrain, which determines the availability of the land for agriculture. The kingdom of Bhutan has a temperate climate with marked local variations: cold winters, cool and warm summers. However, the southern region of Bhutan bordering India is an exception. Here the climatic conditions are typically sub-tropical—a combination of heat and humidity. Because of its location in the sub-Himalayan region, the kingdom receives copious rainfall, mainly during the summer months—1,520 mm. (or 16 inches) in the submontane valleys and 5,080 mm. (or 200 inches) in the southern parts bordering the Duars.

References have been made to the travels of Stephen Cacelle and John Cabral, who arrived in Paro, Bhutan, on March 25, 1627. Barring the visit of these foreigners, no European travellers visited the kingdom. James Renell, the first Surveyor General of India, was deputed to survey the territories in India going to be ruled by the British. Renell mentions the 'Bhoutan Rajah' as the ruler of the territory. Moreover, travels were discouraged because the natives were averse to this idea. They did not allow foreign visitors to their country. Historical records show that George Bogle visited Bhutan as the head of the mission to Tibet in 1774. Warren Hastings was highly inquisitive about the territory of Dev Rajah. Under his instruction Turner visited parts of Bhutan. Among others Dr. William Griffiths, a well-known Botanist, arrived at Punakha, situated in east Bhutan on April 1, 1838. He left behind a short sketch of the natural vegetation, economy and the people.

Reaching Bhutan was always difficult; Himalayas are difficult to cross and the kingdom of Bhutan survived the onslaught of time. It is also important to understand the psyche of the people of Bhutan. European explorers belong to a different category. They visited Bhutan in the guise of pilgrims and collected important information on the territory. On the basis of this information a map of Bhutan and adjoining areas was compiled. The sub-Himalayan states of Sikkim, Bhutan and Tibet continued to attract the attention of the explorers. Bhutan is divided into four dzongdey (administrative zones). Each dzongdey is further divided into 20

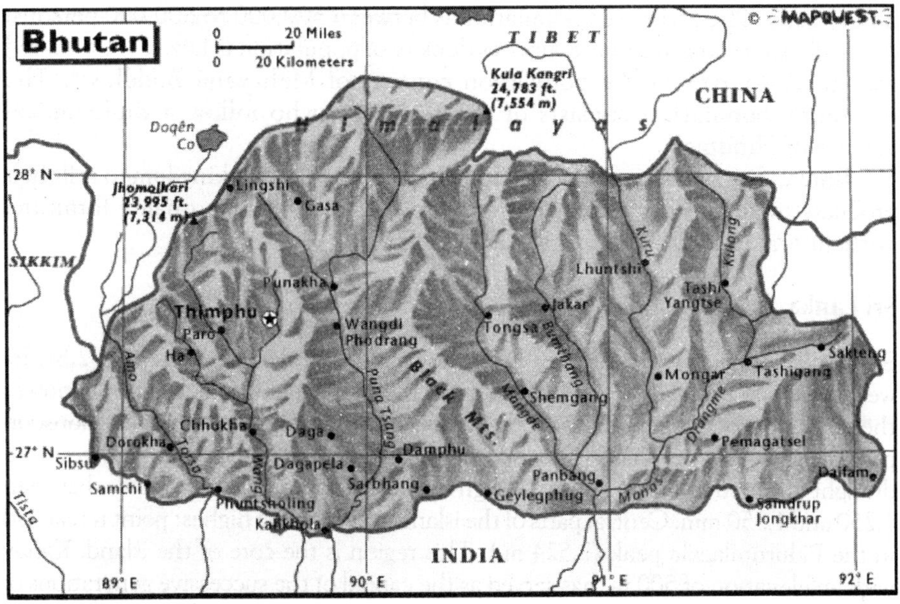

Source: http://atlas.mapquest.com/atlas/

Map 2.7: Map of Bhutan

dzong Khag (districts). The kingdom is administered by a commissioner, locally known as dzongtsap.

The kingdom of Bhutan is an example of a delicate relationship between altitude and location. The southern region, bordering the valley of Assam, experiences humid sub-tropical climate. The next region has a micro-thermal Himalayan climate. This region extends over the Inner Himalayas. Above 20,000 ft. the climate is Tundra type.

A cross-section of the country from south to north shall take one through several vegetation zones. Tropical moist deciduous trees are found in the humid sub-tropical south and temperate coniferous trees in the micro-thermal climatic zone. On the other hand, northern Tundra region has the alpine vegetation. Bhutan's economy is pre-dominantly based on agriculture and the allied activities, such as animal husbandry. The natural environment is harsh. It does not leave much scope for the diversification of the Bhutanese economy. Most of Bhutan is unproductive. Agriculture is mainly concentrated in the narrow Duar strip and the major river valleys of central Bhutan. Breeding and pasturing of cattle is an important source of income. Among the animals, the Indian cow and half-breed yaks are predominant. Cattle raising is also done for dairy products. Agriculture is confined to well-watered valleys and in the warm humid Duars at the foot of the mountains. The main food grain crops include rice and buckwheat. A very coarse variety of barley is also grown in the higher valleys. Potatoes grow up to an altitude of 14,000 ft. In the absence of a regular census, the size of population is an intelligent guess only. But these estimates differ vastly. The estimated population is between 300,000 to 800,000 persons. As in all other mountainous countries density of population is low. It is estimated that three-fourths of the population consists of Mahayana Buddhists. The remaining population consists of the Nepalese, who follow a Brahmanical version of Hinduism.

Bhutan has signed a treaty with India, which enables the kingdom to sell their products to India. The main items of external trade include quality furniture, horticultural products and fruits (Karan 1967: 79).

Sri Lanka

Sri Lanka is one of the northern-most islands of the Indian Ocean (Map 2.8). Its west coast is as rainy as the Malabar Coast of the Indian peninsula, a continuation of the same monsoon regime. Most of the rainfall occurs during the southwest monsoon period. A part of the island is dry and the typical breaks in monsoon may result in drought of varied intensity[17], even though the rainfall amount is not so low—between 1,250 and 1,850 mm. Central parts of the island are hilly. The highest point is reached in the Pidurutalagala peak (2,524 m.). This region is the core of the island. Kandy with an elevation of 500 m. has served as the capital of the successive generations of Sinhalese kings. It continues to be the hub of the Sinhalese culture and religion. However, its congenial climate also attracted the European colonisers, particularly the Dutch and the British. The primeval forestland was cleared for plantation crops, such as coffee, tea, rubber and coconut, which were introduced by the colonisers. The hill ecosystem was disturbed because of the plantation crops in spite of their high export value.

Sri Lanka is a microcosm of the subcontinent—its history, culture, polity and population dynamics have all followed the same path; quest for democracy, balanced development and peace has continued, although with not much of a success. Elections are held time and again. Official attempts to address the problems of civil conflict, have by and large failed. The state policy is not adequate for handling the national problems, such as Tamil insurgency, and underdevelopment. It has suffered setbacks due to the indefinite postponement of the parleys between the Sinhalese and the Tamils.

These groups have not been able to achieve a working understanding among them. India, like a big brother, has tried to help the island nation. But it has also tended to overlook the problem both in the maintenance of democracy as well as cultural harmony. Moreover, India has not been active in resolving the inter-ethnic problems. Not much has been done to bring the fighting factions to a discussion table. Tamil-Sinhalese relations have remained strained all the years. The other side of the story is that the political parties in Sri Lanka do not have unanimous views on this vexed ethnic problem.[18] The centre-state relations in India are such that the federal central government has no major role to play. The government has moved cautiously so far and can take no steps, which are likely to strain the relations with Sri Lanka. No central government in India can afford to ignore the interests of the ruling National Progressive Alliance (NPA)

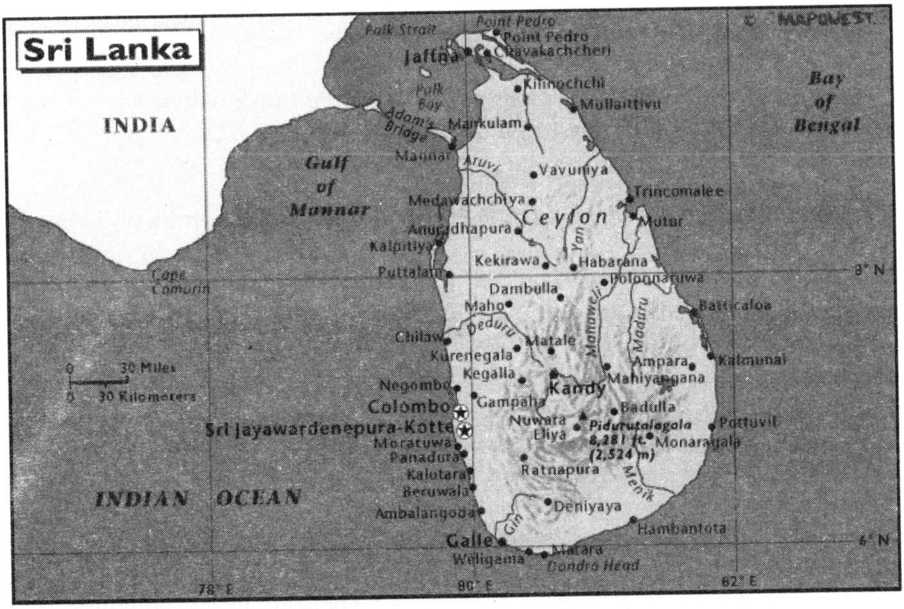

Source: http://atlas.mapquest.com/atlas/

Map 2.8: Map of Sri Lanka

in New Delhi, and the state government ruled by AIADMK in Chennai. Paradoxically, India has not been much vocal on the question of Tamil insurgency. While India is opposed to militancy anywhere in the world, it appears to be non-committed towards Sri Lanka. This task, of peace-keeping, to a very large extent, was taken up by a far-away country—Norway. The Norwegian efforts, laudable by all means have brought peace, but not a viable settlement of the on-going ethnic problem.[19]

The island nation has a shared history with other countries of the subcontinent, particularly India. Tamil, a Dravidian language is spoken by the Tamils of Sri Lanka. On the other hand, Sinhalese belongs to the Indo-European family. Thus the political problem in Sri Lanka today is a matter of adjustment and understanding between the Dravidian and the Indo-Aryan language groups. Tamil is a cultural continuity of the Dravidian peninsular India and the insular Sri Lanka. The Indian government has always offered a helping hand but it is always difficult to differentiate between diplomacy and rhetoric in political speeches of the leaders.[20] Non-alignment does not mean neutrality in situations of violence.

Today, the island economy is mostly dependent on peasant agriculture. However, export-oriented plantation crops have always sustained the nation. Ninety per cent of the foreign exchange comes from the export of spices, tea, rubber, and coconut. The area under food grains, mainly paddy has diminished, as a consequence. Rest of the land is utilised for export crops. But there may not necessarily be a cause-and-effect relationship between the two. The prosperity of the planters is one of the causes of the poverty of the paddy growers, the aborigines known as the Vedides and other marginalised sections of population.

Maldives

The island nation of the Maldives is an integral part of the subcontinent. Formerly known as the Maldives Islands, it is currently known as Divehi *Jamhuriyya* (Republican Maldives). There are 1,087 islands, although only 215 are inhabited. The archipelago stretches 804 km. in a north-south direction and 159 km. east-west. Located in the north-central part of the Indian Ocean, 1,087 islands are clustered in 19 atolls about 643 km. south-west of Sri Lanka and almost the same distance from the southern tip of the Indian Peninsula in Cape Comorin (Map 2.9). The total length of the coastline is about 2,393 km. There is ample scope for fishing, which is the main source of national income. The islands encompass the total area of 298 sq. km. with an estimated population of 238,000 persons. The population of the islands is reported to be growing fast; estimated annual growth rate being 3 per cent. However, the nation is not in a hurry to implement a planning policy on population. Average household size is 5.4 persons. Divehi administration has no schemes on the anvil to regulate the population growth.

Maldives consists of tiny islands just a few feet above the sea level. Secular changes in the sea-level often result in the disappearance of islands. Some disappear while others appear due to the same phenomenon. There are several navigable channels in the east-west direction. Notable among them are Kardivie or Five Degree Channel, Viemendu or Kolemadedula, one-and-a- half Degree Channel and the Equatorial Channel. These channels provide ways for marine navigation. The islands enjoy a typically tropical climate with little seasonal change in temperature. Rains are copious and have a marked seasonality. The southwest monsoon current brings rains during the period from June to August and the northeast monsoon current during the period from November to March. The average annual rainfall is 254 cm. (100 inches). Both the temperatures and rainfall support the vegetation cover, which is by and large, an Equatorial type. Till recently Maldives was a British Protectorate. This status lasted for a period of 78 years from 1887 to 1965. The air base at Gan is the only relic of the British possession. The country is being administered by the constitution adopted in 1953, but reactivated in 1968. It is interesting to note that Maldives does not have a department of defence, post office, ministry of labour, social security administration. It is alleged that elections are held on the basis of universal suffrage.

All population over 21 years of age have a right of franchise. The government is controlled by vested interests. Legal system is based on Islamic *Shariah* modified in the light of the modern law. For example, all draconian laws have been abolished. The administration functions on the basis of 19 *verin*, or a Chief, for each of the 19 atolls. The *verin* is, however, assisted by an elected committee. There is *kateeb* (headman) in each populated atoll, both the *Verin* and the *kateeb* are appointed by the president.

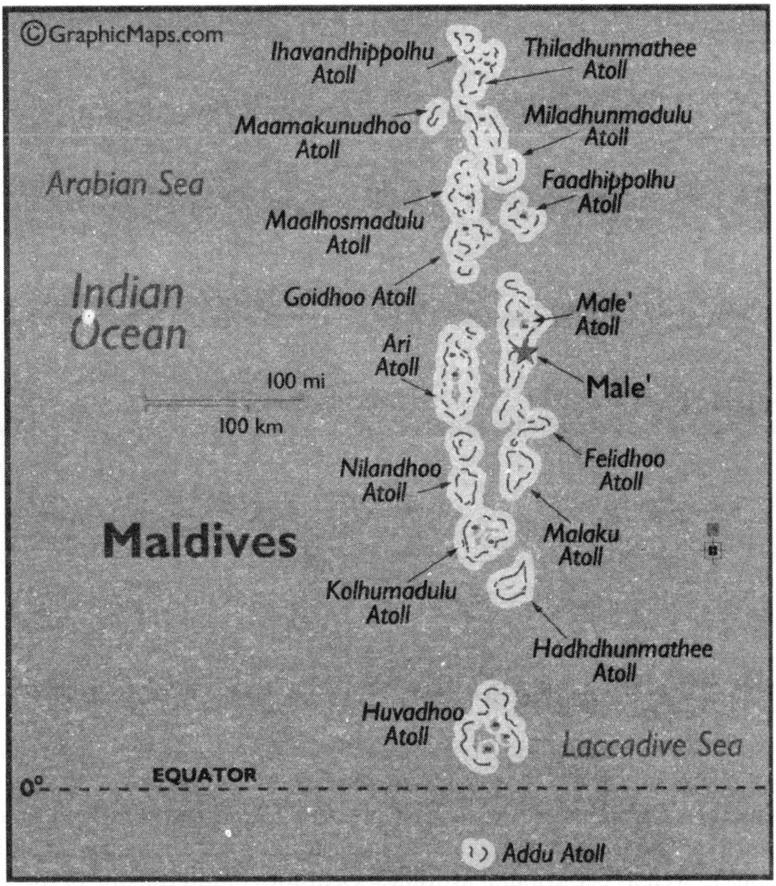

Source: http://www.worldatlas.com/webimage/countrys/asia/mv.htm

Map 2.9: Map of Maldives

A series of development projects have been launched recently to promote fishing and food processing industries. Plans have been chalked out for improving communications, tourism and health services.

India

India, as we know it today, is a residual state in terms of size. It encompasses all the land after excluding peripheral areas, such as Afghanistan, Pakistan, Bangladesh, Nepal, Sri Lanka and Bhutan. The freedom movement culminated in independence in 1947. However, independence came with the Partition of the country. Initially, Pakistan consisted of East and West Wings. The Eastern Wing, however, seceded and a new country, Bangladesh, came into being. From the very beginning, the Eastern Wing of Pakistan was not a viable arrangement politically as well as strategically. Perhaps it was a response to the operation of the centrifugal force in

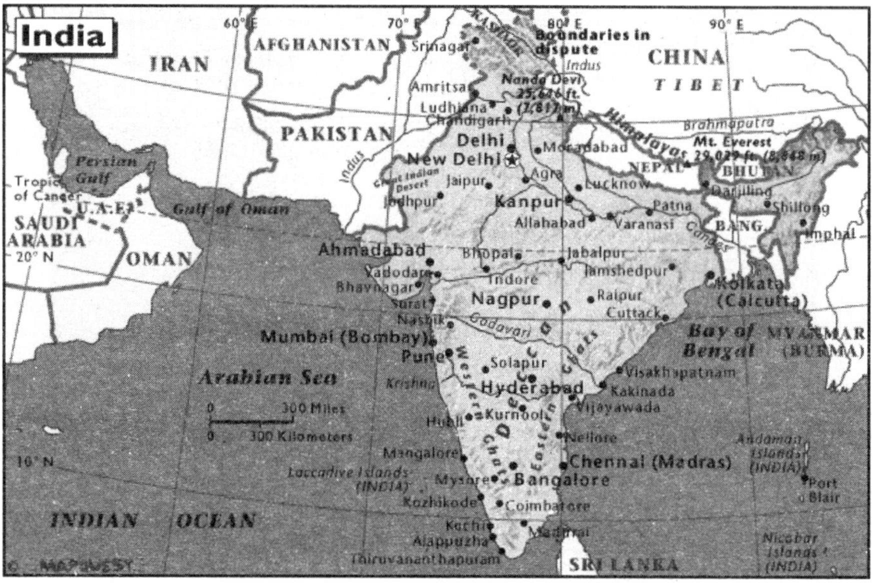

Source: http://atlas.mapquest.com/atlas/

Map 2.10: Map of India

the polity of Pakistan. It was not realised that religion may not be an effective tool/ basis for unification.

It may be posited that the idea of the subcontinent was born in India. Despite the geography and the strong belief in the eastward drift of the cultures and civilisation, today's India is the mother of the subcontinent, the homeland and the centre of diffusion of the major ethnic, magico-religious and the culinary traditions as well as behavioural attitudes. Centuries of intermixing of the people coming from different parts of the world have resulted in diversity in folk tradition, arts, music, dance forms and literature.

To describe the surface features of India is to tread the oft-trodden path. However, a brief outline is necessary here in order to familiarise the reader with geographical and locational specificities. India extends from the Himalayas in the north to the shores of the Indian Ocean in the south (Map 2.10). The Indian peninsula tapers towards the south. The peninsula is skirted by the Arabian Sea on the west and the Bay of Bengal on the east. The coastline extends from the Rann of Kachchh in Gujarat to the mouths of the Ganga in West Bengal. The country encompasses a major chunk of the North Indian Plain. The northwestern part of the Indus valley, *i.e.*, upper reaches of the Satluj and the Beas and the upper stretch of the Indus and its tributaries including Ravi, Chenab and Jhelum are a part of the Indian state.

In climatological terms, the northwestern part of the subcontinent is an arid zone, known as the Thar Desert. It extends over parts of Rajasthan, and the Sind Province of Pakistan, particularly the region lying to the east of the Indus river in Sind. The international boundary between the two countries extends from Kachchh in Gujarat to Ishkuman and Yasin (Kishanganga Valley) in the northwest. In the northeastern sector, the crest of the low hills of Nagaland, Manipur and Mizoram describe the borders between India and Myanmar. The southern part of India consists of a plateau. The peninsular plateau is bordered by the relict mountain ranges known as the Western and the Eastern Ghats. It is because of this location that the country receives an abundance of sunshine and rain. The Ghats have passed through successive phases of a long history. They play an important role in the distribution of monsoon rains. In fact, the Ghats act as a water-divide. Parts of the plateau lie in what is known as the 'rain shadow' area of the Western Ghats. The general pattern of the distribution of rainfall suggests a decreasing tendency from the east to the west. These factors also define the economic opportunities available to the people mostly dependent on agriculture. The region experiences a dry tropical monsoon type of climate. But there are large variations in the distribution of temperature, rainfall and humidity. The northern boundary of the subcontinent lies in the Himalayas. A segment of the international boundary between India and China lies in Jammu and Kashmir, Himachal Pradesh, Uttaranchal, Sikkim and Arunachal Pradesh. Paradoxically, the upper parts of the Siachen glacier, lying in the Karakoram range is in the possession of Pakistan, while India controls the lower parts of the glacier. The northern sub-Himalayan region experiences an alpine type of climate, more particularly in Kashmir as well as Himachal Pradesh and in the newly created state of Uttaranchal. On the other hand, the peninsular region experiences hot and humid climate. Because of the diversity in weather, climate and water supply, a variety of crops are grown in different parts of the country. These crops include wheat, paddy, bajra [a millet], jowar, dorapern, sorghum and maize. Besides, there are oilseeds like groundnut, rape, mustard and sesamum. Sugarcane, cotton, and pulses, like gram and tur are also grown over vast areas of the north Indian plain. Plantation crops, such as coconut, coffee, tea, and rubber are also grown on a large scale. It is noteworthy that agriculture absorbs a very large population, about 70 per cent, who are dependent on peasant agriculture. Agrarian relations have never been smooth. After the abolition of *zamindari*, particularly in Oudh and parts of West Bengal, in the early fifties, a healthy change was noticed. But it was short-lived. There were mechanisms to accumulate land in a few hands in each village. The ruling Congress Party felt the need for land reforms but found it ineffective. Even today, agriculture contributes to the gross national product in no small a way.

India lies in the tropical monsoon type of climate. There are large variations in all elements of climate; rains decrease from east to west, while it increases from west to east in the narrow coastal plains lying on the outer fringe of the peninsular plateau. As one moves from east to west on the North Indian Plain, the dependence on irrigation increases. In regions, where irrigation is not viable, crops are grown with dry farming methods. The large-scale fluctuations in rainfall result in intermittent drought. For centuries altogether, drought has resulted in famines of varied intensity. The history of the country is full of episodes of drought, famines and death. These conditions have often led to migration from dry to humid regions – men and animals moving in large numbers from Gujarat and Rajasthan to wetter regions in the east and the north. The interior of the peninsula is a dry region, perhaps the best example of a rain-shadow area; the Western Ghats are rainier than the interior parts of the peninsula. Winter rains from the northeast current of the monsoon (November to March) compensate to a very large extent, for the deficiency of rains during the southwest monsoon period (June to September). The history is also full of references to the conflicts on sharing the river waters, *e.g.*, sharing the Kaveri waters between Karnataka and Tamil Nadu; or sharing the Narmada waters between Madhya Pradesh and Gujarat. Suggestions have been made in political circles to link the rivers of India on the pattern of national electricity grid. But such suggestions are not viable, simply because there is not much water available in the rivers of the North. It is also not viable because of the costs involved. The north Indian rivers flow in a low lying plain. The land rises gradually from the Ganga delta to the Indo-Gangetic divide in Delhi and Haryana. The plateau rises abruptly south of the Yamuna. It reaches an average height of 600 m. (about 2,000 feet) in a few kilometres. In any case, the cost would be much higher than the benefits accruing to the farmers. Moreover, the plan of diversion of river waters will disturb the ecological balance seriously. The plan of diversion also appears detrimental to the little economies of the tribal and riverbank communities. The people of the states from where the proposed link canals are likely to pass will oppose the idea because it will necessarily lead to redistribution of population, particularly the tribes who practice shifting agriculture (*jhum*). The civil strife and the water riots in the ecologically fragile regions of central India are symptoms of the impending crisis.

One may go back to the early days of India and Pakistan when the dominions signed on the Indus Treaty, which provided for sharing the river waters between India and Pakistan. The proposed diversion of river waters would lead to dangerous consequences. The problem of relations between India and Bangladesh is critically related to the Farraka Barrage and the quantum of water to be stored to flow due to the objections raised by Bangladesh. The issue is not easy to resolve.

India is a vast country in terms of geographical area (3.29 million sq. km.). But there are other large countries, such as Australia, Brazil, United States of America, China, Canada and Russia. The country is about one-third of the size of Canada and a little more than one-third of China. However, it surpasses most of the countries of the world in terms of population size. The country is more populous than the combined population of Russia, United States of America, Canada, Australia and Brazil. The population of India crossed the billion mark several years ago. It is estimated that the population has now reached the mark of 1.02 billion. The comparison with the developed countries of the world is of very little significance in so far as the quality of life is concerned. Nobody can be proud of the size of population simply because of the numbers. The critical indicators of density of population and growth rate reveal more than these numbers. It is true that in terms of population size India is second to China. But China is far ahead of India in terms of human development.

Population is unevenly distributed among the states of the union. Five states of the Indian union, viz. Uttar Pradesh, Bihar, Maharashtra, West Bengal, and Andhra Pradesh share among themselves more than one-half of the country's population. The situation in the south is different. For example, the state of Uttar Pradesh has more people than the combined population of Kerala, Karnataka and Tamil Nadu. Geographers generally relate the size of population with the natural resources and availability of agricultural land. It is considered as the most important indicator of pressure of population on land.

India has been governed by a Constitution adopted in 1952. While the Constitution has been amended several times during the last 54 years, the basic structure of governance has remained the same. There is a dualistic system of governance. The central/union government is in charge of the basic all-India matters. Defence, finance, international trade, transportation and communications are controlled by the central government. There are other departments, such as agriculture, rural and urban development, education, health, excise and sales tax, which are within the jurisdiction of the states. There are twenty-eight states and seven union territories. Each of these states has a legislative assembly and a legislative council. Thus the legislative powers are in the hands of the states. However, there are many areas, such as human resource development (education and health), which are considered as the state subject. There are areas of administration/jurisdiction in which the rights of the union government are shared between the state and the union government.

The Constitution of India envisaged a form of federalism to bind the states with the union. Although the spirit of federalism has not been implemented fully, it continues to be the basic guiding principle. Centre-state relations remain a matter of definition. Larger states take a lion's share out of the central development funds leaving little scope for the smaller states. Indian federalism is based on the socio-cultural theory recognising multiple identities as the basis of collective coexistence. Indian federalism is a political principle necessary to stabilise the pattern of constitutional distribution of power and as an administrative arrangement to legitimise distribution of power and jurisdiction between the federal authority and the state (Khan 1997: vi).

The Indian political tradition has come to terms with the process of nation-building. However, it is quite different from the federal nation-building. 'What is usually referred as nation-building without the prefix federal becomes in a continental polity of India's spatial (geographical size), density of population and diversity of culture and society...an unimagined authoritarian imposition, an uncritical regimentation and a negative processes of duress to change unity and union of a complex and variegated federal situation into mere uniformity and conformism of a simple unitarian state' (Khan 1997: 1).

There is a set of priorities for transformation of a free democratic society with accumulated economic disparities and inter-regional imbalance, in conditions of scarce resources, low technology and mal-development of goods and services into a modern polity. There are five priority areas for national reconstruction:

(i) Stabilising the foundations of participatory democracy;
(ii) Promoting secular polity;
(iii) Ensuring social justice;
(iv) Subordinating anything else to the rule of law; and
(v) Maintaining an open society.

In all such generalisations what is missing is the focus on federal nation-building (Khan 1997: 2). It is customary to recognise the environmental framework to understand the diversity of cultures within the mosaic of ecologically diverse regions. Diversity has been given not much of a weight as compared to the unifying processes integrating the underlined little cultural worlds into all aspiring empires. It is generally accepted that the regional differentiation owes its origin to the ecological setting. The similarities or expressions of uniqueness have lived longer than the empires (Ahmad 1994: 56).

Nehru, the first prime minister of free India, died as a crusader of peace, although frustrated by the failures of the Nehruvian political system both at home and abroad. There were the old phantoms in the new garb—poverty, hunger, underdevelopment, parochialism, regionalism and an ailing economy at home and the strains and stresses of the global political and economic order abroad. Our neighbours—China and Pakistan—had their own compulsions. Nehru could bring about neither peace nor development, which were the age-old dreams of an ordinary Indian. Nehru's *panchsheel* failed to lure China. But his accomplishments were eclipsed by the over-arching threat to national integration, a piece of his own imagination and national security. He died after remaining in office for seventeen years. In the end he came to be known as a dreamer rather than an achiever. His scientific temper and the massive support of the people could not diminish the agony of the man who wanted to undo the legacy of poverty, parochialism and the seemingly disparate levels of regional development.

A Resume

In sum, we can say that the subcontinent presents a case of multiple identities and not of pluralism. There are visible and invisible links, which connect the South Asian nations and define their identity. We can describe the region as having a common heritage in all indicators of historical and contemporary development. It is observed that the British Empire in India played an important role in bringing the disparate parts within the ambit of a single 'whole'. The ethnic scene today is an end-product of centuries of intermixing of racial and ethno-lingual elements. Moreover, the impulses of social, ethno-lingual and economic change are an index of fusion and not fission. As already noted, two boundaries delimit the subcontinent: the Hindu Koh-Himalayas in the north and the Indian Ocean in the south. Geography, more particularly locational factors have contributed to the over-arching unity. The people of the subcontinent are diverse and disparate. Our identity is based on the integrative processes operating through the ages. One notices the coexistence of diverse elements from primitive communism to the liberal democracy and petty capitalism. The people of South Asia have passed through phases of history of division, further sub-division and integrative force was weak. The historical experience is that religion has never been a uniting force. The integrating forces have been ephemeral. There are nomads, wandering in different parts of the subcontinent in the midst of agrarian communities. They live on the periphery of the river valleys. Agricultural produce has sustained the people through the ages. A hierarchical social system has survived, even though the old forms of culture and economies are becoming anachronistic. The population has subsumed diverse forces like caste, ethnicity, language, religion and local tradition. The idea that people have inherited different traditions of society and polity does not mean separation and indifference. The purpose of democracy is to strengthen the individuality of the individual but not the rejection of heterogeneity.

Languages like Sanskrit, Tamil, Persian, Hindi-Urdu and English have operated as a unifying force at different stages of our social history. In modern times, transport and communication systems, trade of commodities, education, urbanisation and industrialisation have played the role of integration. The traditional routes of migration have brought the people in far-off regions into a broader frame of unity. The Early Indians consisted of tribes mostly practicing egalitarian values, and unconscious of any hierarchy. Astro-Asiatic languages, despite the fact that they were surrounded by the developed languages, were not wiped out. Language provides scope for give-and-take across the language domains. Middle India presents a case of overlap of language domains. It also displays evidence of one language dominating over the other. The tribal groups have lived in comparative isolation. The surface features of the subcontinent played a role in defining the overall identity.

We will examine this evidence in some detail in order to understand the phenomenon of division and union. Rapid change has been registered in regions of agricultural development. Industrialisation and urbanisation have promoted the process of integration. But it could not remove the disparities. Even the democratic process has not been able to reject these structures. Indian democracy has, by and large, adapted itself to the social system. If caste is a basic element of identity, the geographical distribution of caste groups will be a factor, which no political party can ignore. Political parties have often been responsible for retaining the *status quo*. It is not the unity but the diversity, which is the hallmark of distinction of the people of the subcontinent.

END NOTES AND REFERENCES

1. Pushtuns/Pakhtuns are described as true Afghans. Besides Afghanistan where they are a major ethnic group, Pushtuns are also found in the North Western Frontier Province of Pakistan. It has been found that an overwhelming proportion—95 per cent—of the Pushtu speakers of Pakistan live in the North Western Frontier Province districts of Bannu, Kohat, Mardan and Peshawar. They are also found in Baluchistan, Sind and the Punjab. They are by and large a trans-humance group and move along familiar routes, spending summers in Afghanistan and winters in Pakistan.
2. Across the Himalayan range lies the Xinjiang province of China; Tarim Basin where the ancient civilisation of Kashgar and Khotan were born. The narrow panhandle of the Wakhan region of Afghanistan served as a shatter belt between the Tsarist and the British Empires in South Asia.
3. Khyber Pass has been a major entry point for travellers between Kabul and Peshawar.
4. Read Hindu Koh, not Hindu Kush. It appears that the term was adopted later. Abul Fazal Allami, author of the Ain-e-Akbari mentions it as Hindu Koh. All major entries into the subcontinent were made via the passes in the Hindu Koh range. The change of the name from Hindu Koh to Hindu Kush is a mystery. But the general opinion is that the invaders who plundered time and again nick named the range as Hindu Kush, meaning the killer of Hindus. The reference is to the hazardous mountain range with high passes frozen during the major part of the year. Several expeditions from India with the intent of crossing the mountain range reported a high toll of life.
5. The traffic consisting of nomads along with their animals who reached Kabul found the Khyber Pass as the main gateway to India. The Kabul river meets the Indus near Attock.
6. The major urban centres of Afghanistan, such as Kabul (1,830 m. or 6,000 ft.), Jalalabad (550 m. or 1,800 ft.) and Qandhar (1,070 m. or 3,500 ft.) lie at different altitudes. Kabul enjoys a pleasant weather throughout the year, comparable to some extent with Shimla or Nainital, well-known hill resorts in the Siwalik Range. Winters are cold and summers are generally pleasant. However, Afghanistan is a dry country with the average annual rainfall not exceeding 305 mm. (about 12 inches). But most of the water supply comes from the snowmelt of the Hindu Koh and other mountains.
7. The partition of the country resulted in the dislocation of population and migration of Muslims from India to Pakistan and the Hindu and Sikh population from Pakistan to India. As is well known, millions of people of either country shifted their residence under the threat of violence that erupted before and after independence.
8. The struggle for the liberation of East Pakistan resulted in a tripartite war: India, Pakistan and the people of East Pakistan. The main issue was that the Eastern Wing of Pakistan was an example of neo-colonialism imposed by the Pakistani regime.
9. The Liberation movement in the East Wing of Pakistan was based on the notion that the Bengalis had a separate identity and cannot be subsumed in the population of West Pakistan.

10. Bangladesh came into being as a separate autonomous and sovereign state at the culmination of a bloody struggle for separation. It was India's full-fledged involvement in the 1971 war that helped the Muslim Bengalis to realise their dream.
11. See the author's interview to the *Indian Currents* (Vol. XV, Issue 27, 2003: 20-21).
12. There is a climatic indicator of dry conditions and measured on the basis of aridity index. The Thar Desert in India and Pakistan is a part of the arid zone, which extends from the Aravallis in India encompassing the western parts of Rajasthan, Sind, Baluchistan, southern Afghanistan to southern Iran and Saudi Arabia.
13. 'Most of the Bengali speaking non-locals merged with the local communities and were integrated therein. The Urdu-speaking and educated upper and middle classes, including professionals and big businessmen, moved from East Pakistan to West Pakistan by air or sea in search of better alternatives. However, the bulk of Urdu-speaking people, largely consisting of illiterate masses, farmers, landless workers, artisans and petty merchants, decided to stay on in East Pakistan. Unfortunately, these refugees mostly originated from the Indian states of east Uttar Pradesh and Bihar. They were concentrated in ethnic enclaves for group solidarity and security which restrained their mobility and reduced the chances of employment' (see A.F.M. Kamaluddin in Kosinski (ed.), *Population Redistribution and Development in South Asia*, Dordecht, Holland, 1985: 224. According to official estimates of the census of 1951, there were 1,18,181 Urdu-speaking refugees. Out of this, 18,819 originated from Uttar Pradesh, 97,349 from Bihar and 2,002 from either Punjab or Delhi, p. 224, *ibid*.
14. A simple definition of secularism is the political validity. There should be evidence to show that all our actions are in harmony with the state policy committed to the values of democracy, failing which either secularism or democracy would remain slogans.
15. The two neighbours India and Nepal have a soft border. The *terai* belt extends parallel to the foot of the Siwaliks in Uttar Pradesh (India) and Nepal. Visa is not a condition but visitors from either side have to prove their identity.
16. The on-going civil strife is largely due to the hold of the landed peasantry on the state machinery. One may apprehend that the present struggle being waged by the Maoists is going to be a long-drawn battle.
17. The monsoon type of climate of both Sri Lanka and India is characterised by its erratic behaviour. Fluctuations in the rainfall amount and its intensity lead to a sequence of floods followed by drought. If the drought continues for more than three weeks it hits the paddy crop. Drought continues to be a major hazard to agriculture in the subcontinent.
18. Tamil struggle for recognition as a national minority identity is not new. Central governments in India have time and again made appeals to the two parties, the Tamils and the federal governments of Sri Lanka. However, the ground reality is that the state government in Chennai (formerly Madras) will be required to concur with any policy measures taken by the government of India.
19. There is a history of the Tamil problem in Sri Lanka. In fact, the differences between the two opposing factions have their roots in the cultural and linguistic tradition. The struggle is between the two diverse groups—Dravidian Tamils and the Indo-Aryan Sinhalese.

20. There is a continuing crisis of leadership in the politics of Sri Lanka. The clash between the two parties, one, represented by the president of the country and the other, represented by the Prime Minister is a symptom of political factionalism. This divergence of opinion between the two national political parties disturbs the peace process. The Sri Lankan establishment has to make a firm and honest effort to resolve the problem of militancy and the causes responsible for it.

3

Geography Behind History

In this chapter, we have tried to describe the physical framework of the South Asian subcontinent. An attempt has been made to recognize the major physiographic regions. The underlying assumption is that the geographical factors determine the boundaries between the regions and within the regions. The evolution of the state from ancient India to the modern times bas been analysed. Different streams of ethnic and ethno-lingual groups came to occupy the subcontinent at different points of time and that set the stage for ethnic intermixing par excellence. The chapter also contains the material on region formation and their geographical extent and boundaries. Two types of regions have been identified: (a) regions based on physiography; and (b) regions based on agro-climatology. There are references to regionalism and regional consciousness of the people of the subcontinent in modern history.

The assumption is that geography sets the stage on which the human drama is enacted. It is this terrestrial space on which patterns emerge indicating the on-going process of interaction between nature and the humankind. A general description of geography of the subcontinent is to serve as the foundation for historical exploration (Sastri 1981: 34).

The basic assumption of this statement appears to be a re-incarnation of the by-now discredited theory of environmental determinism.[1] Human initiatives notwithstanding the broader framework of development in the human society is a result of the location on terrestrial space[2] and the associated factors that come along with it, e.g., distance from the sea, altitude, climate, accessibility by land and sea, social organisation, and more importantly, the organisation of production forces. Possibilism does not negate the influence of environmental factors on human life.[3] It is the package of the natural resources, and the human cognition of this potential, that the interactive process begins. Each of these factors operate individually as well as in combination with one another. They define the choices available to the humankind within a certain framework of environment and social organisation at a given level of technology.[4]

Environment is a comprehensive term. It plays a determining role in influencing the conditions of the humankind on this planet Earth. The geographical surroundings of places and regions offer unlimited possibilities for human development, materialistic as well as meta-materialistic, *i.e.*, enabling, or not enabling, the humans to select a given path of development at a given point of time in a given segment of space. The humans adopt a set of technology to utilise the development potential as perceived by them.[5]

Surface Features

What is the nature of the land surface of the subcontinent? We may examine the role of the geographical factors, such as location, altitude, lay of the land and the elements of climate in determining the course of South Asian history. The raised relief map of Central Asia shows a series of mountain ranges with awe-inspiring heights.[6] Inter-montane valleys separate the mountain ranges. There is a sequence in which the mountains and the valleys occur. Mountain ranges follow the inter-montane valleys, which follow the mountain ranges, which follow the inter-montane valleys and the out-skirting hill ranges. The mountains of the subcontinent are a wall-like structure, which define boldly the northern borders with Tibet, a province of China, and the Central Asian states. The mountain ranges have always restricted overland movement from China to India and *vice versa* to a bare minimum. All human movements have taken place through the difficult mountain passes along the Himalayas from Indus to Brahmaputra, as well as from Indus to the Helmand river in Afghanistan. However, a thin traffic of nomads along with their animal stock has always managed to cross these passes, howsoever difficult they might have been. This reveals the protective role of the Hindu-Koh Himalayan chain of mountains in the history of the subcontinent. It is because of the Himalayas that the Indians have been able to retain their distinctive ethnic/racial traits. There may be other reasons as well, but terrain and accessibility do matter. This is not to deny the fact that both India and China have been influenced by each other's cultural traits, ethnicity as well as linguistic affinity. It may be noted that this influence remained confined to the northern border regions from Baltistan and Ladakh in the northwest to Arunachal Pradesh, Nagaland, Manipur, Mizoram, Meghalaya and Tripura in the northeast. The evidence, mostly ethnographic and linguistic, is scattered all over the region. This marginal interaction notwithstanding the mountain ramparts have helped the people of the subcontinent retain their uniqueness.[7]

Thus it is evident that the partially enclosed character of the South Asian subcontinent owes its origin to the two outstanding features, which delimit the region—the Hindu-Koh Himalayan Mountains in the north and the Indian Ocean in the south.[8] In the north, the bordering rim of the lofty mountains has an impressive cluster of the highest peaks in the world.[9] The mountains remain snow-bound for the greater part of the year. In fact, they have acted as a formidable barrier throughout our history. However, the mountains have never been regarded as impregnable. There are passes all along the mountain chain from Afghanistan to Assam and beyond. It is through these passes that a small traffic of nomads, pilgrims, *sanyasis* and the Boudh *Bhikshus* has always moved. Even the armies have crossed these passes to reach the subcontinent. For example, in 1962 the

Chinese army moved into Arunachal Pradesh in the northeast and Ladakh in the northwest, after crossing difficult passes. Therefore, the mountain ranges have played a critically important role in defining the ethnic and cultural characteristics of the South Asian subcontinent. It was in this context that the famous Urdu poet, Iqbal, described the Himalayas as a vigilant sentinel guarding our territory.[10]

Towards the south, the Himalayas lose their height in the Siwaliks and finally, give way to the vast alluvial basin—the North Indian Plain. The geographical location of the Plain is such that it promotes a feeling of security, self-sufficiency and a psychic feeling of exclusiveness. The Plain extends from the Punjab to the Ganga delta in Bengal and Bangladesh. Punjab Plains have been built by the five rivers, Satluj, Beas, Ravi, Chenab and Jhelum, which have deposited the detrital materials over the geological ages. The joint stream of these five rivers, the Panjnad, pours their waters in the Indus. The Plain reappears in the Brahmaputra valley in Assam. The Plain has been described as a gable (Wheeler 1964). The shorter wing slopes towards the Arabian Sea in the Plain of Indus and the longer wing slopes towards the Bay of Bengal. The mighty Brahmaputra, locally known as Jumna in Bangladesh, joins the Ganga, and the united stream flows into the sea. Jumna, Meghna, Padma and the Hooghly rivers flow through a deltaic tract, with an imperceptible slope, before they pour their waters in the Bay of Bengal.

On the south, the North Indian Plain gives way to the Deccan Plateau, which rises gradually, but at some points it rises abruptly and looks like an escarpment. The Plateau reaches its maximum height in the Nilgiris, where the highest peak is named as Doda Betta (2,636 m.). The Ghats consist of a series of hill ranges, such as the Sahyadri, Ajanta, Nilgiris, Cardamom, Palni and the Anaimalai Hills. The Central Vindhyan Complex has a series of mountains extending from west to east, such as the Vindhyachal, Satpura, Mahadeo, Maikala, Bhanrer and the Kaimur Range. There are several erosion surfaces, like Chhotanagpur, Chhattisgarh and the Bastar plateau. The tip of the Peninsula dips into the Indian Ocean at Cape Comorin (Kanyakumari). To the east-southeast of the Peninsula lies the island of Ceylon (now Sri Lanka) the Sarandeep of the medieval geographers and travellers. The Palk Strait separates Sri Lanka from the mainland. The subcontinent resurfaces in Sri Lanka where it reaches the maximum height in Pidurutalagala (2,524 m.).

The Vindhyas are a complex mountainous tract. The relief on its northern face is rugged and not very negotiable. The southern face, on the other hand, is marked by a gentle slope. On the south, the land declines abruptly with a general escarpment like relief, sloping down to the Narmada basin. The Narmada basin is bounded by an escarpment, which stands like a wall. The Satpura, Mahadeo and further eastward Maikala ranges dominate the scene. Further eastward, the Tapi river flows on the south of the Satpura roughly parallel to the Narmada river. This complex of hills and escarpments is generally recognised as the dividing line between north and south India (Wadia 1976: 406). However, these mountains have never been impregnable. Inter-communication has never been hindered. The give-and-take between the northern and the southern regions of India has continued through the ages.

It may be recapitulated that the two delimiting factors—mountains in the north and the sea in the south—have contributed immensely to the uniqueness of South Asia as a result of the intermingling of the different racial and cultural strands. These factors have operated through the ages. In fact, they have acted in a way so as to make the South Asian subcontinent a unique cultural realm, a melting pot for the mixing of diverse peoples and diverse cultures. The mixing of the racial and ethno-lingual groups has made the subcontinent a museum of social and cultural history.

The peninsula tapers towards the south, finally disappearing in Cape Comorin. From the tip of the peninsula, the coastal plains—Malabar in the west and the Coromandel in the east—run in the northwest-northeast direction respectively. All major seaports like Cochin, Goa and Mumbai are located on the Malabar Coast. They have provided anchorage for ships from Africa and Mediterranean lands.

The Himalayan mountains are arranged in ranges one after the other. The inter-montane space has accommodated little culture worlds from Gilgit, Hunza and Baltistan to Arunachal Pradesh. With all its detailed topography these cultures have survived the ages of historical development. Their own evolutionary history is not properly documented.

Looking at the historical links that India established with the neighbouring countries across the nation, it may be concluded that the Satvahanas, Pallavas and the Cholas established contacts with the eastern world—the Malay Peninsula and Indonesia. It was due to their interest in trade that the country emerged as a maritime power. Their glorious tradition of seafaring and shipping was known all over the old world.

The peninsula has a triangular shape made up of very old rocks from Satmala-Ajanta ranges to the Nilgiris, the plateau on its western edge is very steep. On the other hand, the plateau has a gentle fall on the east. This complex of hills is defined as the Eastern Ghats. Between the Eastern Ghats and the Coromandel Coast lies Carnatic lowland. The eastern coastal plain is wider than the western coastal plain. If you look at the Western Ghats from the west, they look like a gigantic fall often rising in steps from the shore line, hence the name ghats. The average elevation of the plateau is more than 600m. from the mean sea level. In the northern part, the Ghats rise more than 1,200 m. The altitude increases towards the south culminating in the Nilgris. The continuity of the Western Ghats is broken by the Palghat Gap. Its north-south length is about 32 km. People have traversed this gap time and again.

The basaltic lava plateau is covered by a thick layer of the lava deposited here during the Vindhyan times. Geologists describe it as the Deccan Trap. The weathering action has shaped the surface suitable for citadels and fortresses in the midst of the hills. They have locationally strategic value. During the Maratha period, the mountain formations helped establish their military supremacy. The plateau is covered by the black soil, a residual of the weathered lava sheet, considered fertile for cotton production. A dense forest has covered the seaward side of the plateau. There are several passes in the Western Ghats, which have enabled human movement. Reference may be made to Trimbak pass near the source of Godavari. Pimpre pass provides routes from Sopora and Kalyan to Nasik. There is yet another pass known as the Nana pass which provides access to Junnar and the Konkan. It was on the mountain spurs that the fortifications, such as Bhimsankar and Chakan are located. Another pass is known as the Bhor Ghat or Khandala pass with a height of about 610 m. The Amba pass connects Ratnagiri with Kolhapur. There are other important passes such as the Shencottah Ghat and the Aramboli pass. Long-range changes in the level of Coromandel coast have been quite common. Once upon a time, the commercial cities known as Korkai and Kayal were flourishing in commercial activities. The cities, however, are buried under the sea. Similar changes have been recorded in the Gulf of Cambay and other neighbouring areas.

Roundabouts and Cul-de-Sacs

It is necessary to examine the salient features of structure and relief of the subcontinent, which have determined the space relations. The tract of land between Peshawar (Purushapura in Sanskrit) now in Pakistan, and Jalalkot, now known as Jalalabad in Afghanistan, is negotiable through the historic Khyber Pass. There are, of course, other passes in the northwestern mountain complex linking it with Afghanistan. Likewise, there are passes in Baluchistan, which have provided access to the nomads as well as to the warriors. It is believed that the Greeks, Greco-Bactrians, Huns, Kushanas and the Indo-Iranians (Aryans) came by the same route. Further north, the passes in the Hindu Koh range, particularly in the Bamiyan province of Afghanistan, viz., Kowtal-e-shebar, Kowtal-e-salang and Kowtal-e-anjaman, the upper streams of the Kabul river and its tributary, the Kunar river, provide natural routes linking the northwestern region of the subcontinent with the northern slopes of the Hindu Koh range and the regions lying beyond. This segment of the subcontinent has been the main theatre of intermixing of the people of diverse ethnic origins. The low-lying valleys of northern Afghanistan have always attracted the pastoral nomads. In some valleys, traditional agriculture has also been a viable alternative to nomadic pastoralism. Further east-northeast, the city of Mazar-e-Sharif is the outpost of the seat of power based in the Kabul valley. There are tiny little streams, like the Zhob, Kurram and Loralai, to mention only a few, with cascaded valleys on the western slopes of the Sulaiman Range. These valleys have a thin alluvial veneer suited to traditional agriculture. The region has sufficient archaeological evidence to show that the early communities, familiar with the art of agriculture settled in the tiny villages during the period 4-5 millennia B.C. These valleys have also served as routes for the adventurists and the warriors wandering between Baluchistan and southern Iran. The Oxus (Amu Darya) has been crossed time and again by travellers from Mazar-e-Sharif to Termez, in Uzbekistan and *vice versa*. However, the traffic has never been one-way. The medieval historians recognised the Oxus as the natural boundary between Afghanistan and trans-Oxania (the regions lying beyond the Oxus, *Mawara-un-nahar*). This area has also been described as eastern Turkistan or Turan.[11] As noted earlier, the Indus receives the waters of the Kabul river and its tributaries near Attock. It may be recalled that the Mughal emperor Akbar (1556-1605 A.C.), built a fort at Attock. He perceived this place as a strategic point for the defence of his vast empire in South Asia (Box 3.1). It was the cis-Indus region on the east bank of the river in Punjab, which had a great potential for agriculture, the real economic base on which the Mughals were crucially dependent.

Box 3.1: The Road to Kabul

From Torkhan to Jalalabad, the road runs through a broad corridor between two parallel snow-caped ranges, the Safed-Koh on the south and the Hindu-Koh on the north. Beyond Jalalabad this corridor splits into the two valleys of the Kabul river and its right bank tributary, the Surkhab, a prudent road steers clear of the Kabul river's gorges. Even the modern road is prudent enough to prefer the Khyber hump to the Warsek slot—and anyway, since the completion of the Warsek dam, this particular gorge has been filled to the brim by the waters of a reservoir, leaving no room at all for a road to find a passage between lake-edge and mountainside. Above Jalalabad the ancient road prudently chooses to give the Kabul river the widest possible berth, though it was heading for a city through which the Kabul river flows, and from which it takes its name. The ancient road swerved away up the Surkhab valley, making for the Khurd Kabul Pass; and here, at Nimla, the Mughals had an imperial garden where they camped when they were travelling to Kabul from Delhi at a leisurely bullocks' and elephants' pace. The modern road rashly chooses to hug the right bank of the Kabul river all the way up to Kabul city; and immediately above the confluence of the Kabul river with the Surkhab, a short stretch of gorge, taking road and river by the throat, gives a foretaste of what lies ahead. Then the valley opens out again and, as quickly, contracts into another gorge that seemed interminable on Wednesday the 20th April 1960 (Toynbee 1961: 47-78).

East of Indus, travellers intending to reach the North Indian Plain have often preferred the northern route running parallel to the base of the Siwaliks, the foothills of the Himalayas. Further north, the movement is obstructed because of the hilly terrain. On the south lies the Thar Desert, which has never been considered as a viable alternative route. But there is the example of Humayun, the son of the Mughal emperor Babar, who under unfavourable circumstances crossed the Thar desert and reached Umarkot, in the Tharparkar district of Sind, on his way to Iran to seek asylum in that country. The northern route takes one to the river Yamuna, at the point where the river debouches from the hills, near Yamunanagar, in Haryana. It may be recalled that the nomads, warriors and the stray travellers coming from the west and the northwestern parts of Asia crossed the Indus river frequently.[12]

An alternative route lies further south, which connects Multan with Delhi via Hansi, a small town, now in Haryana. The northern route takes one to the upper Ganga-Yamuna *doab*, one of the most fertile tracts in the subcontinent. The region has always remained exposed to the traffic of nomads, wanderers and the armies of the enterprising invaders originating from parts of western and central Asia. But its location has been a great disadvantage to the local people. It is not easy to describe the misery of the people living in this 'corridor of intrusion'. They were plundered time and again by the intruders. Perhaps their history was written in blood and erased by water.

The North Indian Plain stretches for thousands of kilometres in the east-southeast direction between the Indus in the northwest and the deltaic Bengal-Bangladesh in the east. Moreover, the plain has its eastern-most segment in the Brahmaputra valley in Assam. On the south, the plain merges with the peninsular plateau almost imperceptibly. Geologists inform us that the northern flank of the peninsular plateau in southern Uttar Pradesh and Bihar lies buried under a thick layer of the alluvium deposited by the rivers, such as the Chambal, Betwa, Ken and the Sone, all originating in the mid-Indian plateau region and emptying themselves in either the Yamuna or the Ganga (Ganges). Historians tell us that the North Indian Plain along with the northern segment of the peninsular plateau constituted the core of the Madhyadesa long back in time, perhaps in the Puranic period (c. 600 B.C.—600 A.C.). It was the hub of the Delhi Sultanate in the medieval period (1200 A.C. and beyond). The puranic Madhyadesa extended from the banks of the legendary Saraswati (perhaps the Ghaggar river)[13] of northern Rajasthan in the west to the Rajmahal hills in the east and from the Siwaliks in the north to Narmada (earlier Narbada) in the south. The region came to be known as Hindustan.

Confusion on the question of defining the extent and boundaries of the region notwithstanding, Hindustan lives in the public memory. In real language terms, all parts of northern and central India where the main language is Hindi or Urdu came to be known as Hindustan and the spoken language as Hindustani. The lingua franca changes from Hindi-Urdu to Punjabi in the northwest, to Bengali and Oriya in the east and to Gujarati and Marathi in the southwest of this region. The Hindustani-speaking region ends with the Rajmahal Hills in Bihar. The North Indians are called Hindustanis in Bengal and Assam as well as in the Deccan region, just as all South Indians in North India are described as Madrasis (after Madras, now Chennai). North Bengal is a transition zone between Hindustani and Bengali language realms on the one hand, and the Bengali and the Assamese on the other. However, there are also minor language groups, such as the Gorkhas and the Bodos. This is a transitional zone to the Assamese in the eastern and the northeastern regions.

Language historians believed that the Maithili was the mother of all the three major languages of eastern India—The Bengali, Assamese and Oriya. A general view is that Hindustan does not have a single core or a single periphery. Cores have shifted as the battle for controlling Hindustan has continued over centuries. As cores have shifted, peripheries have also shifted. In fact, there have been multiple cores and multiple peripheries. For example, places, as far away from each other as Kanyakubja (now Kannauj) and Purushapura (now Peshawar) have served as capital cities. These were the two hubs from where the King Emperor Harsh Vardhana ruled over the subcontinent. Then there was the Delhi Sultanate, which had its nerve centre in Delhi. The Sultanate encompassed the entire region of the

Indo-Gangetic divide, upper and middle Ganga-Yamuna *doab* and the trans-Ganga region of Rohilkhand in western Uttar Pradesh. Perhaps the Delhi Sultans were not capable of controlling the territory lying further east. The eastern region was a part of the Sharqui kingdom, based in Jaunpur in eastern Uttar Pradesh. The kings of Gaur ruled over the deltaic region of the Ganga in West Bengal and Bangladesh. There were, as the historical facts reveal, many gaps in the so-called no-man's land in between. They were the centres of bandit power.

The valley of Brahmaputra in Assam was another core region, although not as extensive as the other regions mentioned above. The surrounding rim of mountains and hills from Arunachal Pradesh to Tripura constitute the Assamese core. The Garo-Khasi and Jaintia Hills, the outer Himalayan range in Arunachal Pradesh and the curvilinear hills in Nagaland, Manipur, Mizoram and Tripura together form the eastern frontier zone of the British Empire in India up to independence in 1947. It may also be noted that there was no viable political boundary between this region and the neighbouring territory of Burma (now Myanmar). There are disputes between India and China on the northern boundary of Arunachal Pradesh. A part of the territory is claimed by China. There is confusion on the demarcation of the McMahon Line, which is yet to be accepted as a boundary between the two countries (Box 3.2).

Box 3.2: The McMahon Line

The boundary in the eastern sector is 1,140 kms. long and runs from the eastern limit of Bhutan to a point near the Talu Pass at the tri-junction of India, Tibet and Burma. This line is called McMahon Line after Henry McMahon, a British representative who signed the 1913-14 Simla convention. The boundary was established along the Himalayan crest of the northern watershed of the Brahmaputra, except where the Lohit, Dihang, Subansiri and Kemang rivers break through that watershed. The only variance from the watershed principle is near Miayetun and the two Tibetan pilgrim places Tstokaro and Tsari Sarpa.... In the eastern sector, China claims about 94,700 sq. kms. of Indian territory, occupying 19,000 sq. kms. including the Kameng Frontier Division, the Subansiri Frontier Division, the Siang Division and three-fourths of the Lohit Division of Arunachal Pradesh. In 1911, Tibet revolted against the Manchu Dynasty and until the end of 1912 there were continuous skirmishes on the Sino-Tibetan Frontier. In 1913, Tibet declared full independence.... After full discussion, the three plenipotentiaries (Chinese, Tibetan and British) initiated the map and drew up the draft convention.... The McMahon Line then, did not create a new border but merely confirmed the natural, traditional, ethnic, and administrative boundary existing between India and Tibet—a boundary which has been recognised for centuries (Sukhwal 1985: 263-64).

It shows how difficult the task of clearing the historical backlog is. It is here that India and China meet. In fact, Myanmar constituted a part of the Raj, until it was given a special status in the 1935 Act of India as a separate country. The India-Myanmar boundary does not correspond to any historically established criteria. The *janajatis*, such as the Nagas, Kuki-Chins, Mizos and the Tripuris, to mention only a few, are living on both sides of the boundary. At present the Indo-Myanmar boundary is well defined, and the more important thing is that the two concerned countries accept the position on the ground. But within India the boundary between Manipur and Nagaland has given rise to civil strife, with the Nagas demanding the constitution of another state, which according to them should include Nagaland as well as parts of Manipur state where Nagas are predominant. The Naga youths have been pressing for this reorganisation. The on-going agitation has disturbed the normal life due to the blockade of all movements on roads. The Manipur state is virtually cut off from the rest of the country.

The rivers have acted as natural boundaries throughout human history. The Mughals, for example, recognised the channels of the five rivers of the Punjab as boundaries between the Sarkars. The *doabs* between the two confluent rivers emerged as an acumen for the human settlement and for the pursuit of agriculture.

Within the subcontinent, boundaries have been delimited rather arbitrarily. The boundary between India and Pakistan, for example, does not follow any river channel or a range of hills. As it happens, history and practice are the main criteria in the delimitation of boundaries. The boundary of the princely states of Bikaner and Marwar with Sind, Khairpur, and Bahawalpur in Pakistan, became the Indo-Pakistan boundary in this sector. In the British Punjab (undivided Punjab) none of the tributary rivers of the Indus system served as natural breaks for boundary formation. Similarly, the line of control (LOC) in Jammu and Kashmir simply shows the position on the ground, perhaps, as we believe, the situation is likely to remain the same unless it is recognized as an international boundary between the two countries. This may not be the ideal solution. But the recent political history cannot be ignored. So far, neither of the two sides have agreed to convert the armistice line into an international boundary. The Kashmiris are certainly opposed to any such move for the partition of the state.

The Republic of India, Bharatavarsha of the ancient period, as we know it today, is a distinct geographical entity occupying the central place in the South Asian subcontinent (Ali 1966). It is the largest country in the region in terms of area, population and resources. It is a country of great geographical extent and its people are of diverse ethnic composition, dialects, languages and the folk arts and crafts. In fact, its vastness in size, diversity in the natural environment and in material cultures is amazing. So is the disparity in human development. All these factors add to its uniqueness. It extends from the snow-covered peaks of the Himalayas in the north to the Indian Ocean in the south. The Bay of Bengal and the Arabian Sea have a coastline extending for thousands of kilometres from the state of West Bengal to the Rann of Kachchh in Gujarat. Taking Delhi as the centre, the Indian territory stretches on all sides encompassing the vast expanse of the North Indian Plain from the upper Indus valley in the west to the Brahmaputra valley in the east. The encircling line of hills around the Assam valley and the Shillong plateau together constitute the international boundary between India and Bangladesh. The tropical sun and the monsoon regime are the two elements that together exercise an influence on the life of the teeming millions of South

Asia, who are dependent on agriculture and the rural artisan industry. As we know, the Himalayas constitute the boundary between India and China. In the following paragraphs, we will describe the macro regions of the subcontinent based on physiography.

Regions and Regionalism

Historians, historical geographers and archaeologists have noted that the regions based on the surface relief features and the drainage basins of rivers have acted as break points between one region and the other. They also believe that the dualism between the early river valley settlers and the hillsmen is based on the geographical framework. Often the boundary between the two is knife-edged. However, these surface variations have caused diversity in the formation of cultural regions and their evolutionary history (Appendix II).

Macro Regions

These regions have been identified on the basis of physical features. Physiography is defined as a branch of geology, which deals with the development of the existing features of the terrestrial space. The surface features are the product of a combined action of the external forces of the lithosphere. Physiography deals with topography and the landscape produced by the weathering agents. These agents may be identified as running water, glaciers, wind and the sea waves. These forces operate on the global scale. But the base on which these weathering agents operate takes its form and shape due to the earth-movements. The higher the relief, the more intensive is the erosional work of these agents. The term surface features apply to various kinds of landscapes in mountains, plateau surfaces and the plains. The South-Asian subcontinent consists of the following macro regions:

(a) The Hindu-Koh-Himalayan Mountains
(b) The North Indian Plain
(c) The Peninsular Plateau

The Hindu-Koh-Himalayan mountain complex consists of a series of mountain ranges. Across the Great Himalayan Range lie the Karakoram, Zanskar, Ladakh and the Kailasa ranges and the plateau.[14] To the south of the mountains lies a vast alluvial plain, the North Indian Plain. On its outskirts in the northeast lies the Brahmaputra valley in Assam. In the northwest beyond the Indus lies the Kabul valley, an alluvial plain built by the Kabul river and its tributary streams. However, there is no comparison between the Kabul valley and the Assam valley. They differ both in the depth of the alluvium as well as in its width and density of the detrital deposits. Moreover, Kabul valley is distinguished for its glacial past and its altitude. On the southern outskirts of the North Indian Plain lies the Peninsular Plateau. The geological view is that the alluvial deposit of the Plain, in its eastern and western segments, has covered the northern flank of the Plateau. The alluvial deposit laid down by the Ganga and its numerous distributaries and the Brahmaputra river in the deltaic region have broken the continuity of the Plateau, with the result that the original form of the plateau lies buried under a thick layer of the alluvium. The outlying segments of the Plateau re-emerge in the Meghalaya Plateau, popularly known as Garo, Khasi and Jaintia Hills, and Kachar, Tripura and Chittagong Hills. The Plateau also re-emerges in the west and the northwest. The Aravallis and the Potwar Plateau are the relicts of an earlier phase of geological history.

The northwestern flank of the Plateau is a dry region known as the Thar Desert. The Thar extends from the Aravallis to the Indus and beyond. In fact, the Thar is a part of a vast arid zone, which extends from Rajasthan to Sind and from Sind to southern Baluchistan. The dry region also encompasses the south-western parts of Afghanistan, southern Iran and the whole of North Africa, up to the Atlantic seaboard. The Indus breaks the continuity of the desert and flows through it in Sind. The dry zone of the southwestern parts of Afghanistan is a similar case—the Helmand river flows through it and breaks the continuity of the desert. The Helmand is one of the major inland streams of the world. However, the comparison ends there. Helmand is worlds apart from the Indus in terms of its size, length and the volume of water it carries. The northeastern periphery of the subcontinent is delimited by the Himalayas in Arunachal Pradesh. The largely denuded hills of Nagaland, Manipur, and Mizoram delimit the region and separate it from Myanmar.

Can we consider coastal plain as a regional unit of the first order? There is a difference of opinion among geographers on the question of considering the coastal plain lying to the west of the Western Ghats and to the east of the Eastern Ghats as part of the Peninsular Plateau, or it is a physical division of the first order in itself. The Eastern Ghats is a series of broken hills running parallel to the coast. It is not as imposing as the Western Ghats. However, it is customary to recognise the coastal plains as a part of the Plateau. As noted earlier, the Island of Sri Lanka is an extension of the Indian Plateau, even though it is a separate country.

Meso Regions

Each of the three-macro divisions of the subcontinent described above consists of several meso regions, or second-order regions. In the first place, each meso region has a broad uniformity in its physical characteristics, such as relief, geomorphological history, drainage, climate, natural vegetation, animal life and soils. Although at the micro level these regions may have differences in the natural environment and/or superficial features, there is a broad unity in their general appearance, which distinguishes them from the other regional units in the neighbourhood. The North Indian Plain, for example, has a broad uniformity in relief and geomorphological characteristics. But at the lower level there are striking differences in all these respects. Brahmaputra valley and the Punjab Plains are both alluvial plains, but they sharply differ in details due to geographical location and the microclimates. However, local details have to be ignored in order to identify the unity at the next higher level of generalisation. The natural history, geomorphological features, climate and the soil are not always distributed in a way so as to correspond perfectly with each other. They often overlap. The regional boundaries are, therefore, generalised. The transition from one region to the other is not a line but a zone through which one type of region changes into the other.

The Hindu-Koh-Himalayas have striking differences in their layout and location. For example, the Eastern Himalayas lie closest to the sea. The Hindu-Koh, on the other hand, lies farther away from the sea. However, due to their location, the mountains are exposed to the tropical sun and humidity. Similarly, the Assam and Kashmir Himalayas are both part of the northern mountain complex, but they differ in many respects, particularly in their glacial features, distribution of precipitation, character of natural vegetation and the associated wild life. Likewise, the North Indian Plain has a broad uniformity from the Hooghly and Tista rivers in the east to the Indus in the west. But even a common observer may notice the important differences in relief and climate between the Ganga delta and the delta of the Indus. The Punjab plain and the north Bihar plain have differences in their micro relief. These differences are caused by the depositional work of rivers, local climates, natural vegetation cover and the soil. Even the difference between the plains of western Uttar Pradesh and eastern Uttar Pradesh is quite well marked, although the climatic regime is by and large the same. Evidently, the criteria for the identification of the second-order regions (meso-level regions) are not necessarily the same as for the delimitation of the first order regions.

The concept of regions is hierarchical, which means that all macro regions have meso regions and all meso regions are divisible in micro regions. An analysis of the regional contents reveals that all regions exist at all three levels at the same time. The classification is based on a scale defining the hierarchy. At the macro level, it is usual to recognise the first order regions on a general basis, e.g., a physiographic division at the highest level of generalisation. This generalised classification defines the basis of their identity.

As pointed out earlier, the narrow coastal plain—Malabar and the Coromandel coasts, situated on the outskirts of the peninsular plateau, is recognised as a region of the first-order. But there is no consensus among the geographers on this suggestion. Information is not sufficient to recognise the coastal plain as a first-order region. Geologists believe that the coastal plain is an integral part of the peninsular plateau itself and cannot be identified as a separate region.

Regions of the Hindu-Koh-Himalayan Mountain Complex

(a) Kabul River Valley
(b) Herat, south of the Paropimisus Range and the Firoz Koh
(c) Qandahar
(d) Balkh and Badakhshan, north of the Hindu-Koh Range bordering Turkmenistan
(e) Helmand Desert
(f) Trans-Hindu-Koh Region
(g) Waziristan, Chitral, Yasin, Gilgit and Hunza

(h) Karakoram, Baltistan, Kargil, Dras and Ladakh
(i) Kashmir Valley
(j) Upper Chenab Valley, Kishtwar and Doda
(k) cis-Pir Panjal Range: Jammu, Punch and Rajauri
(l) Himachal, upper reaches of the Beas, Spiti and the Chenab rivers, south of the Satluj and between Satluj and Beas
(m) Uttaranchal: Garhwal, and Kumaon Himalayas
(n) Nepal, Sikkim and Bhutan
(o) Eastern Himalayas
(p) Arunachal: Great Himalayas, Abor, Miri, Dafla, and Mishmi Hills
(q) Purvanchal Hills, Naga Hills, Barail Range, Aimatol Range, East Manipur Hills, Lushai Hills, and Tripura Hills

As already noted the Hindu-Koh-Himalayas are a division of the first-order. Both the mountain ranges are considered as young in age in geological terms. Geologists hold the view that like the Alps they are young-folded mountains. However, the agents of denudation, such as rivers and glaciers have largely modified the original relief of the mountain complex. There are several identical characteristics. Among these features, the more pronounced are high altitude, high degree of slope and the extremely rugged character of the terrain. The mountains have several altitudinal zones. Each zone is characterised by variations in vegetation cover, water supply and soil formation. The differences in climate and natural vegetation are caused by differences in altitude, aspect and the soil type. By aspect, we mean the direction in which the mountain range is arranged, north or south. Aspect is an important factor. The natural vegetation, snow-cover and the duration of sunshine are all dependent on it. Aspects of the Himalayas, which face the south, are far more sunny and rainy. They receive lashing monsoon rains leading to the growth of dense natural vegetation. However, aspect itself is not sufficient, much depends on the altitude. Both climate and natural vegetation and the associated animal life have an interesting altitudinal zonation. These diversities within the Himalayan Hindu-Koh mountain complex have resulted in a number of second order regions, which can be identified easily.

The Kashmir valley, which lies in the northwestern region of the subcontinent, belongs to the Himalayan realm. It is surrounded by high mountain ranges in the western Himalayas. It has a distinct regional character based on the eco-system of an alluvial plain. The population consists of ethnic and ethno-lingual strands. The climate varies from temperate to alpine type, with a variety of natural vegetation from subtropical to Alpine. The specific ecological conditions, level relief and abundant water supply from snowmelt are some of the distinguishing features of the valley of Kashmir. The surrounding mountain ranges, such as the Pir Panjal and the Great Himalayan Range, have differences in climate and forest types caused by differences in altitude and exposure to the sun and the rain.

In the north, the Great Himalayan Range gives way to the imposing Karakoram Range and the old pre-Himalayan Plateau surfaces, such as Baltistan, Aksai Chin and Deosai, etc. Across the Great Himalayan Range and to the northeast of the valley of Kashmir lies Ladakh, which has a distinct climate marked by low precipitation and cold weather conditions. Ladakh, a good example of a cold desert, along with Kargil and Dras valleys have accommodated distinctive ethnic types. The cold weather season is rigorous. The mercury may drop to minus 40 degrees Celsius in winters. The transition from the valley of Kashmir, a distinctive ethnic region, to the Jammu region is through the Pir Panjal/Pir Pantsal range. Many of the passes in the mountains remain frozen for the greater part of the year. As stated earlier, the traditional routes to Kashmir passed through the Baramula gorge, now under the control of Pakistan. However, India needed an all-weather road to reach the valley of Kashmir. This resulted in the construction of an all-weather road that passes through a tunnel below the Banihal pass. The tunnel provides a safe entry into the Kashmir valley. But the route is not free from landslides, which often block the passage to the valley.

To the east of the Satluj gorge, the mountain range is known as Dhauladhar, which lies in an east-west direction. Further eastwards, the mountains extend into the Kumaon and Uttarakhand regions of Uttaranchal. The general relief, however, is not much different. The Himalayas in this sector are different from the Kashmir Himalayas in rainfall and natural vegetation. Both Kullu and Kangra valleys have a distinct physical appearance; so are the people. The higher rainfall has its effect on the nature of forest types, which are sub-tropical, particularly in the low-lying areas. The general appearance in Kumaon remains by and large the same.

The high mountains extend further eastwards in the Himalayan kingdom of Nepal. There is a broad similarity in the relief features. Nepal has three main physiographic divisions: the Great Himalayas, the Middle Himalayas, locally known as the Mahabharata range, and the Outer Himalayas, or the Siwalik range. The plain area on the foot of the Siwaliks is known as the *terai* belt. The Kathmandu valley is a small alluvial basin lying between the Great Himalayas and the Mahabharata range. The major concentration of population is seen in the Kathmandu valley and the *terai* region. Sikkim and Bhutan are also located in the lap of the Himalayas. They have a characteristically mountainous relief; the climate ranges from cold to Alpine conditions. It is to be noted that the highest peaks of the Himalayan Range, such as Mount Everest and Kanchenjunga lie in this part of the Himalayas. The segment of the Himalayan range east of Sikkim is also described as the Eastern Himalayas.

The Eastern Himalayas are different from the Western and the Middle Himalayas largely because of their layout and placement. It is customary to identify them as a separate sub-region. The change occurs between 86 degrees and 88 degrees East Longitude. To the east of the 88 degrees East, the mountains are exposed to the strong monsoon currents originating from the Bay of Bengal. As a result, the mountains receive a far higher amount of rainfall. The luxuriant growth of the Tropical Wet Evergreen Forest type sharply distinguishes them from the western and the middle segments of the Himalayas. These general features continue without any noticeable change eastwards up to Arunachal Pradesh, which is the eastern-most state of India.

The Indo-Myanmar Hills represent the southward extension of the Himalayan mountain range. However, they have a relatively low altitude and less imposing relief. They rarely rise above 2,000 m. The rainfall in this region is scanty. The forest cover is thick and generally dominated by vegetation types ranging from Tropical Evergreen to the Monsoon Deciduous varieties.

North Indian Plain

The North Indian Plain consists of four segments: the Indus Plain, including the Kabul river valley, the Indo-Gangetic-divide, Ganga Plain and the Brahmaputra Plain. The plain is monotonous in terms of relief features. There are minor local variations. Climate and depositional history have contributed to the lower order differences in relief. Both climate and geomorphology work in combination with each other. They can modify the work of rivers: erosion, transportation and deposition. The North Indian Plain lies parallel to the axis of the Outer Himalayas from west to east. It is interposed between the mountains of the north and the plateau in the south. The Plain consists of the following second order regions:

(a) Sind Plain from Sukkur to the mouths of the Indus near Karachi
(b) Punjab Plain from Indus to Yamuna
(c) Indo-Gangetic-Divide region between Satluj and Yamuna in Haryana and Delhi
(d) Ganga Plain from Yamuna to Rajmahal Hills
(e) Delta of the Ganga from Purnia to the mouths of the Ganga and Brahmaputra
(f) Assam Valley/Brahmaputra Plain from Sadia to Duars

Lying in the extreme west, the Punjab Plain is part of the larger plain of the Indus river. However, a small segment of the Punjab Plain is in India. The major segment of the plain extends from Sialkot to the mouths of the Indus in Pakistan. The southern segment of the plain in Bahawalpur and Multan has an arid climatic regime. It is characterised by a generally dry climate, showing a clear climatic transition between the arid and the wet tracts. The Punjab Plain is divided into several interfluvial plains (*doabs*), such as Bari *doab* between Beas and Ravi, Bist *doab* between Beas and Satluj, Rachna *doab* between Ravi and Chenab and Chaj *doab* between Chenab and Jhelum. The inter-regional differences, as we understand them, appear to have been caused by the shifting river courses. Its semi-arid climate and a dry thorny forest-cover differentiate the plain from the sub-humid Ganga Plain lying to the east of the Yamuna.

The Indo-Gangetic Divide, as the name suggests, is a watershed between the Indus and the Ganga rivers. The divide extends in a northeast-southwest direction, mainly in Haryana between the Yamuna and the Satluj. The divide has a generally alluvial character, but the Aravalli outcrops rise above the level of the land adding diversity to the relief features. The longitudinal valleys between the Aravalli outcrops are intensively cultivated. As in geomorphological character so in climate, the Indo-Gangetic Divide is a transition zone from the sub-humid to the semi-arid types of climate.

The Ganga Plain is an extensive area of generally uniform relief consisting of level alluvial plains. The Ganga and its numerous tributaries deposited a deep layer of alluvium. However, this uniformity is not so pronounced in climate. The rainfall decreases towards the west from West Bengal and Bangladesh to Delhi, Haryana and the Punjab. With this decreasing trend in rainfall, a corresponding change occurs in the natural vegetation and the soil cover. This change reflects the physiographic characteristics of the Upper Ganga Plain and the Lower Ganga Plain. The climate undergoes a significant change as one moves eastwards from Allahabad to Rajmahal Hills. Beyond the Rajmahal Hills lies the delta of the Ganga and the Brahmaputra in West Bengal and Bangladesh. The deltaic character of the alluvial plain is evident in the intricate maze of rivers and their distributaries.

Towards the east, the Plain has another physiographic sub-division in the clearly marked out Brahmaputra valley in Assam. The valley is enclosed by the foothills of the Himalayas on the north bordering Bhutan. They extend up to the Purvanchal Hills. The Shillong Plateau lies to the south of the Assam valley. The region receives heavy rainfall from the southwest monsoon current, with the Shillong Plateau receiving the highest rainfall. Meghalaya and the regions beyond receive prolific rainfall and are characterised by high humidity. These conditions have resulted in a thick growth of natural vegetation.

The Peninsular Plateau

The Peninsular Plateau has remained exposed to weathering since the early phases in its geological history. It is characterised by a broad uniformity in relief features. Its relief has acquired distinctive features of an erosion surface. There are, however, local variations in the geomorphological features caused by ontogenetic forces. It is believed that the differences in relief were caused by the differential uplift of the plateau. Climate and the successive cycles of erosion have affected the process of soil formation, both in terms of depth and soil profile. These differences have resulted in the diversity of natural vegetation. In geological terms, the Peninsular Plateau is described as a 'horst' which means a solid crust-block or a shield which has remained a stable landmass of great rigidity and has been unaffected by any folding movement. The differences in altitude and location have caused local variations in climate, soil and the vegetation cover. Geologists agree that the Plateau has reached the stage of senility. In fact, successive cycles of erosion and the long-range changes in the sea-level have further contributed to the ageing process of the plateau.

The Peninsular Plateau is divided into the following second-order regions:

(a) Aravalli Hills and the Vindhyan Uplands of east Rajasthan and adjoining Madhya Pradesh
(b) The Thar Desert, west of the Aravallis up to the Indus in Sind
(c) Khandesh: Satpura, Mahadeo Hills, Maikala Range

(d) Chhotanagpur Plateau
(e) Bastar Plateau, Chhattisgarh and the Eastern Ghats in Orissa
(f) Meghalaya Plateau comprising Garo, Khasi and Jaintia Hills
(g) Kachchh and Kathiawar Peninsula
(h) Gujarat Plains comprising Mahi and Sabarmati Valleys
(i) Western Ghats from Satpura Range to the southern tip of the Peninsula
(j) Konkan Coastal Plain from Maharashtra to Goa
(k) Malabar Coastal Plain from Goa to the tip of the Peninsula
(l) Wainganga and Mahanadi River Basins
(m) Deccan Lava Plateau
(n) Karnataka Plateau
(o) Telengana Plateau
(p) Eastern Ghats
(q) Deltaic Tract of Orissa, Mahanadi, Brahmani and Baitarani Rivers
(r) Andhra Coastal Plain and the Deltaic Tract of Krishna and Godavari Rivers
(s) Tamil Nadu Plateau
(t) Coromandel Coast from Krishna Delta to the southern tip of the Peninsula
(u) Central Highlands of Sri Lanka
(v) Coastal Plains of Sri Lanka

The northern flank of the peninsular plateau bordering the Indo-Gangetic Plain consists of a series of tablelands separated by the troughs of the rivers. To the west of the Aravallis lies the Thar Desert. It is an expanse of sand dunes and the 'relict' hills. The sand has accumulated during ages of denudation under extremely hot and dry climatic conditions. The vegetation cover is scanty. The Thar extends from the Aravallis to the Indus river in Sind. In the east, the Aravalli hills separate the Thar from the Vindhyan Upland and the gneissic country. The Aravalli range is quite pronounced near Udaipur, where the highest point is reached in Gurushikhar peak (1,722 m.). The altitude decreases in a east-northeast direction. The western slope of the Aravallis in Udaipur and Banswara is fairly rainy and forested. North of Ajmer, the Aravallis are generally devoid of forest cover due to overgrazing. From Ajmer to Delhi, the Aravallis are divided into several parallel ridges separated by longitudinal valleys. These valleys have sufficient trapped moisture, which is replenished by rain. It supports dry farming crops.

The Vindhyan Uplands consist of the Malwa Plateau and the Bundelkhand gneissic country. A line of scarps and the hill ranges, viz., the Vindhyan, Bhanver and the Kaimur runs eastwards from Vindhyachal mountains to the Chhotanagpur plateau. The Satpura range lies between the Narmada and Tapi rivers. South of the Tapi trough lie the Sahayadiris. East of Burhanpur, Govilgarh Hills merge into the Mahadeo range. The uplands are highly dissected and the soil cover is generally shallow. On the other hand, the river basins have a relatively thick soil cover and are important for their agricultural potential. The middle Indian region, south of the Narmada river consists of a series of erosion surfaces. The Tapi trough has alluvial deposits partly contributed by Purna, a tributary of the Tapi. The historically well-known region of Khandesh lies in this alluvial basin interposed between the Ajanta Hills and the Satpura Range.

The Chhotanagpur Plateau, east of the Son river is characterised by a variety of relief features. The Plateau comprises several erosion surfaces at different levels of altitude. The plateau receives high rainfall, which has resulted in a thick growth of the moist deciduous forest. As stated earlier, the Peninsular Plateau east of the Rajmahal Hills lies buried under the thick alluvial deposit of the Ganga. The Garo, Khasi and Jaintia Hills collectively known as the Shillong Plateau are an outlying segment of the Deccan Plateau. The plateau is highly dissected as successive cycles of erosion have resulted in a complex maze of hills. The representative natural vegetation consists of a variety of tropical wet evergreen forest.

The Peninsular Plateau on its western flank in Kathiawar and Kachchh has lava formations believed to be an extension of the Deccan trap. The Kathiawar Peninsula gives way to the mud and salt of the Rann of Kachchh. The Indo-Pakistan border lies in this sector. The general elevation of Kathiawar is below 200 m., but there are several hill ranges, particularly the Gir range, which rise higher than the surrounding land. The diversity in geological structure and relief has given rise to variations in the soil and natural vegetation, which is mostly deciduous. The Sabarmati and Mahi rivers have built a level plain generally referred to as the Gujarat plain. The west-flowing rivers, such as Narmada and Tapi flow through this trough. They have deposited a thick layer of alluvium giving the region a general appearance of an alluvial basin. One observes a clear transition between the humid west coast and the arid and the semi-arid zone of Rajasthan and Gujarat. The coastal plain lying to the west of the Western Ghats consists of several segments, such as the Konkan, Goa and Malabar. While the Konkan coast is narrow, the Goa and Kanara coast gradually merge into the Malabar coast in Kerala. The outlying scarps of the Western Ghats overshadow the narrow coastal plain. Climatically, it is a hot and humid region with a long rainy season. The coastal plain widens southwards into Kerala, a region with a distinctive character. The Malabar coast receives the highest rainfall from the southwest monsoon current. It has resulted in a thick cover of natural vegetation. These factors, along with the lagoons add to the scenic beauty of the Malabar coast unparalleled in the subcontinent.

The Western Ghats delimit the Deccan Lava Plateau. The Ghats, however, continue southwards into Karnataka and beyond. The average height is between 900 and 1,100 m. The Ghats differ in their physical appearance rather abruptly at a point near Goa. The highly rugged topography of the lava plateau is replaced by the smoothly rounded hills composed of granites and gneisses. The Ghats are interspersed by gaps, such as the Palghat, which has enabled cross-communication between the plateau and the sea. The evergreen and deciduous varieties of forest have covered this segment of the Western Ghats.

The Deccan Plateau is covered with lava deposits over extensive areas. The lava sheets were laid down in horizontal beds during the last phase of the volcanic activity in the Cretaceous period of the geological history of the peninsula (Wadia

1976: 275).[15] These flat tablelands are bordered by hill ranges, such as the Ajanta and Satmala hills. The Ghats, of course, act as a barrier to the Arabian Sea current of the southwest monsoon. The rainfall suddenly decreases towards the east. It is a typical example of the rain-shadow effect. This area is characterised by the black soil, known for its cotton cultivation. The soil is a product of weathering of the lava rock formations over long geological periods. The lava of the Deccan Plateau is replaced by gneisses and granites over the Karnataka Plateau. It is an area of generally uniform relief with elevation ranging between 450 and 800 m. The general uniformity in relief features is, however, disturbed by variations in local climate, natural vegetation and the soil cover. The two main sub-regions of Malnad and Maidan stand out as clear examples of this diversity.

To its east and the northeast, Deccan Lava Plateau is replaced by a series of undulating plains and low-lying areas, such as the basins of the Wainganga and the Mahanadi rivers in Chhattisgarh. There are, however, striking differences in rainfall distribution and forest cover. The Wainganga valley is rainier than the upper Mahanadi basin. The characteristic vegetation consists of the *sal* forest. Among the two, the *sal* forest is common in the former while the teak forest is predominant in the latter. To the southeast of the Deccan Lava country, lies Telengana, a low plateau highly denuded and dissected. The isolated rock features called *monadnocks*, produce some variety in the otherwise open and monotonous topography. The northern part of Telengana basin has scanty natural vegetation. The southern part is mainly an expanse of tropical grasses of the Savannah type. On the south, the Palghat gap in the Western Ghats gives way to the most rugged relief features of the peninsula, like the Nilgiris, Anaimalais, and the Palani-Cardamom group. They have typical 'horst' topography. Besides their rugged relief, these hills are further distinguished on the basis of a rich forest cover, particularly the teak and the sandalwood trees.

As noted earlier the peninsular interior is bordered on the east by a discontinuous line of hills called the Eastern Ghats. They have, however, no comparison with the Western Ghats. The hills consist of three main groups: (i) the northern hills, which lie between Jamshedpur and the Godavari river; (ii) the hills lying between the Godavari river and the Palkonda range and formed mostly of the rocks of the Cuddappah system; and (iii) Tamil Nadu hills lying between the Palar and the Kaveri rivers. Generally these hills receive scanty rainfall, which is unevenly distributed. As a result, the natural vegetation cover is also sparse. These differences are mainly due to its location. The northern hills are more forested than the southern. The Cuddappah hill ranges are generally wooded but not as thick as the northern hills. The Tamil Nadu hills, on the other hand, have some forest growth on their eastern slopes.

The relief features of the mid-Indian region deserve our attention because a very large tribal population is concentrated in this area. They live in the midst of, by and large, untouched nature. The tribal habitat lies in these scarps and hills. The tribal groups, such as the Gonds have carved out their enclaves of concentration between the lines of hill ranges and the river basins. They are the denizens of the land. They are an example of the marginalised masses of the country.

The east coast of India has three main segments—Orissa coast, Andhra coast and Coromandel coast of Tamil Nadu. Mahanadi and the Brahmani rivers have built a vast deltaic plain in Orissa. The southwest and the northeast monsoon regimes have their transitional zone in the deltas of Godavari and Krishna rivers. The rainfall regime of the Coromandel coast is determined by the northeast monsoon. The entire eastern coastal plain and the deltaic tracts of Krishna, Godavari and the Kaveri rivers have been used for intensive agriculture through the ages. Both monsoon currents, *i.e.*, southwest and the northeast, bring rain to the east coast. The Tamil Nadu coast remains by and large dry during the southwest monsoon period, when the entire country receives monsoon rain.

The Islands

The Islands of the subcontinent consist of three major groups:

(a) Lakshadweep Group
(b) Andaman and Nicobar Islands Group
(c) Maldives

The Lakshadweep group lies in the Arabian Sea off the Kerala coast, while the Andaman and the Nicobar group is located in the Bay of Bengal. The Nicobar group extends towards the south till the Sumatra Island (Indonesia) is reached. Maldives is a sovereign country. The island consists of coral atolls. Bangladesh also claims some of the smaller islands off the Bay of Bengal coast but the issue is disputed. There are significant differences among the three groups of islands in structure, relief, population composition and ethnic structure. Lakshadweep group consists of thirty small islands; only ten of which are inhabited. They are coral islands and have a humid, tropical climate. Their scenic beauty attracts tourists from all over the world. Geographical proximity to Kerala has exercised an abiding influence on the culture and the language of the people. In fact, Lakshadweep group can be described as an extension of Kerala. The population has Malayali elements. They are mostly Muslim and the entire population of the island is recognised as a scheduled tribe.

The Andaman and Nicobar Islands, on the other hand, extend southwards as an archipelago off the Cape Negaris in Myanmar. The group extends up to the Indonesian Island of Sumatra. There are two sub-groups—Andamans and the Nicobars. The former has three clusters—North Andamans, Middle Andamans and the South Andamans. Likewise, the Nicobars have three clusters—Car Nicobar, Little Nicobar and the Great Nicobar. The islands are thickly forested and provide sustenance to a small population of *Janjatis*, such as the Onges, Andamanese and the Nicobarese. However, over time, the immigrants from the mainland of India have outnumbered the local tribes. Recently an earthquake shook the whole region of the Indian Ocean leaving behind a trail of devastation and death due to the high wave, described as the Tsunami, hitting the region.

Many of the former prisoners—freedom fighters—who were detained during the Raj in a cellular jail at Port Blair—stayed on with their families after independence. The tropical climate marked by Equatorial conditions supports a dense growth of natural vegetation and wild life, which adds to the scenic beauty of the islands. The Nicobar tribes speak a dialect known as Nicobarese, a branch of Mon-Khmer, a sub-family of the Austro-Asiatic languages.

This survey of the surface features of relief indicates the fact that the river valleys have supported kingdoms whose strength lies in their potential for agriculture. On the other hand, the inter-fluvial spaces are marked by spurs with thin soil cover and lack of level land for agriculture.

A Resume

This chapter presents evidence to show that geographical factors play an important role in determining the course of history. It is argued here that the socio-cultural differentiation among the regions and between the regions in South Asia is based on physiography—the term being used here to imply that relief, structure, drainage, climate, particularly the distribution, intensity and timing of rainfall, length of the growing season, flora, fauna and the soil have a combined impact on the direction of development and regional consciousness. Accessibility keeps the regions either enabling them to interact or else leaving them in a state of dormancy. Philosophically, one may argue that regions, small or large, are not static or cut off from the rest of the world. It is the interactive model that defines whether they can live in constant touch or in parts or complete seclusion. Time and space are inseparable; it is a continuum. One may posit that the changes coming with time unfold themselves in space; that space is irrelevant without the horizon of time in view. At the same time, as evidence suggests, the pace of temporal change is constantly affected by spatial factors and diversity in environmental factors has a major role to play in defining the personality of a region as a whole or in parts. Take, for example, the case of Punjab, both Indian and Pakistani Punjab, where the circumstances were conducive to rapid change, may be compared with, say Uttaranchal or Himachal Pradesh. Centuries have passed without recording any significant change in the life of the people, their thinking, attitudes and the worldview. In such *cul-de-sacs*, Himachal, Uttaranchal or the mid-Indian region of rugged relief, underdevelopment gives an impression of timelessness.

It is necessary to distinguish between material culture and higher culture. Religious ideas, rites and rituals are, by and large, intrinsically related to the material conditions. They consist of the tools and implements, devices, and work related to the seasonality of the cropping practices. The basic assumption is that material culture is an expression of human adaptation to the forces of nature. The methods of dealing with nature change constantly. But the gaps in understanding are not bridged easily. It depends on the human perception of the natural endowment. All social groups/communities are not equally conscious of the potential. Or, we can say, that they are not equipped with, suitable technology, to appropriate the potential. The subcontinent has baffling diversity in culture and the other associated phenomena. It may be posited that the cultural differentiation is a product of accessibility and the knowledge of the diverse ecologies within the subcontinent. It is generally observed that the social groups, communities, with differences in the technological know-how live at different levels of social evolution. In sum, we can say, that location in the terrestrial space is an important instrument of change. Knowledge of the ecology of the terrestrial space is the key to success. We do not know exactly that the resource endowment is unevenly distributed in space. The old statement that lack of knowledge hampers progress is still valid.

In this chapter we have tried to argue that the subcontinent is a curiosity of sorts. The region displays a baffling diversity in population composition: ethnicity, physical appearance of the people and their life-styles, belief systems and the cultural attributes. The dialects spoken in the different parts of the subcontinent are in hundreds. Over and above there are *Janajatis* and *Jatis* with different ethnic antecedents. The caste system continues to be a major determinant of social-interactive behaviour of the people. The people have an unflinching faith in democracy. But the social rank of an individual is determined by the *Jati* he/she belongs to. The governments of the subcontinent have always tended to set higher goals in economic and social development, but a creamy layer at the top appropriates most of the benefits of development. Even democracy has been forced to conform to an unchanging *Jati* system. Elections to the legislative bodies at the centre and the states reveal, more than anything else that the voters belong to different social strata. Very often the voting decision is not individual but collective. The main conclusion is that the political behaviour of the people is controlled, guided (or even misguided). It is not conducive to the emergence of a liberal all-India class.

It has been argued that the social history of the subcontinent is a product of the geographical location. In the early phases of history, the northern region of the subcontinent was blocked by the Hindu-Koh-Himalayan Mountains and the south was blocked by the sea. Both these factors have contributed to the Hindu psyche, a sense of exclusivity, contentment, self-sufficiency and autonomy in decision-making. The general impression is that as a people we are 'unique' and not comparable with any other people in the world either today or at any other time in history. But it is a debatable issue.

END NOTES AND REFERENCES

1. A classical geographical view is that the environment determines the course of development of the human society. Ms. Churchil Semple, an American student of Ratzel, further advanced the idea. She said that man 'is a product of the earth's surface', 'the child of the earth', 'dust of her dust'. Earth has 'entered into his bone, and tissue, into his mind and soul'. Ms. Semple went far ahead of her teacher. Her treatise, '*Influences of Geographic Environment*' (1911) treats all development in human society as a product of the geographical factors. E. Huntington, the noted American geographer, stressed the need for understanding the role of climatic determinism in the progress of civilisation (*Civilisation and Climate*, 1915). Radical geographers like

Richard Peet think that the theoretical base of this theory 'was conceptually weak'. We know that the human society in India faces the consequences of both environmental determinism as well as social determinism.

2. The surface of the earth is a familiar phrase in the geographical lexicon. It is also expressed as terrestrial space. It is the outer-most layer of the earth, where the planet earth meets the atmosphere. The spatial interaction between man and nature takes place on this terrestrial space. American geographer, E.L. Ullman, is credited for coining the phrase 'spatial interaction'. The term indicates interdependence of geographical areas (E.L. Ullman 1954, *Geography as Spatial Interaction*).
3. The philosophical interpretation of man-nature interaction was a contribution of the French School of human geography. The term possibilism was, however, used so as to indicate that man has sufficient scope for different options. The French scholar, Vidal de la Blache, was among those who initiated the concept of possibilism. It developed more as a result of exchanges between geographers, particularly, la Blache, Durkheim and Ratzel. 'Possiblism could still legitimately be regarded as a qualification rather than the negation of environmental determinism' (Derek Gregory, *Dictionary of Human Geography*, 1981, p. 270). The historian Lucien Febvre endorsed Blache's view that 'there are not necessities, but everywhere possibilities, and man as a master of the possibilities is the judge of their use' (A *Geographical Introduction to History*).
4. Technology is a comprehensive term. It refers to the package of methods and tools available to the human society at any stage of social evolution. Technology may be locally developed where it is *in situ*; it can be transferred from one region to the other. Flow of technology depends on the social institutions governing the life and choices of ordinary people. They may enable or may not enable a section of the society to avail a certain technology, even though it is available.
5. All regions possess a natural resource endowment, which may remain unutilised due to the unequal access to the right kind of technology in a certain historical context. For example, landless labourers in the rural areas of the subcontinent are not in a position to utilise the irrigation water available to the village. For them, irrigation is by and large irrelevant. However, they can benefit from it because more labourers would be needed to manage the irrigated agriculture.
6. The Himalayan mountains are arranged in a series: the first are the foothills Known as the Siwaliks, the second is known as the Middle Himalayas and the third, Outer Himalayas or the Great Himalayas.
7. It is established that the Hindu-Koh-Himalayan Mountains have contributed to the uniqueness of the South Asian cultural realm. This would mean a partially enclosed subcontinent. If there were no Himalayas, the northern region of the subcontinent would have easily acquired the Chinese cultural traits.
8. The sea in the south of the subcontinent has also played a role in connecting people. But before the sailing boats, the ocean acted as a barrier in human interaction. It is common knowledge that the Indian cultural tradition reached the Southeast Asian countries, such as Malaysia, Indonesia, and Thailand through the sea. The Indian cultural tradition has been absorbed by these countries, despite the fact that the people converted to Islam long ago.
9. All major peaks, such as the Mt. Everest (8,848 m.), K2, Karakoram Range (8,611 m.), and Kanchenjunga (8,598 m.) are located in this zone.

10. In the words of the famous Urdu poet of the subcontinent, Iqbal, the Himalayas are our doorkeepers and stand like a sentinel. They protect us from any kind of danger lurking on our northern frontiers. His actual words are '*woh santri hamara, woh pasban hamara*'.
11. The region was called Trans-Oxania or *Mavara-un-nahar*.
12. Toynbee, Arnold, J. (1961): *Between Oxus and Jumna*, pp. 47-48. '...and the Kabul river shone out like silver on any night before it dived, behind my right shoulder, into the mouth of the Warsek gorge. Now Kabul itself was as good as in my pocket. But, at this moment of unguarded explications, the river god gave me a warning. The road suddenly broke off shot, and two lines of heaped-up stones guided us into a shingle-bed down which at least five separate torrents were coursing on their bay to join the Kabul river from the Safed Koh. The 'lorry' was a powerful creature, but it was also ponderous and bulky. It was in fact, an outsize version of one of these delivery vans that block the streets of London ...'.
13. Perhaps the reference is to the Ghagghar river, which is a dry riverbed in the northern district of Bikaner and Ganganagar districts of Rajasthan. The dry channel of the river, in years of good rainfall may flow further and enter into the Bahawalpur region of Pakistan where it is known as Hakra, and then if more water is there to carry on, enters Sind where its local name is Eastern Nara (Ahmad 1962, *The Human Geography of the Indian Desert*, unpublished Ph.D. Thesis).
14. The Middle Himalayas in Nepal are known as the Mahabharata Range.
15. 'Towards the close of the Cretaceous period, subsequent to the deposition of the Bagh and the Lameta beds, a large part of the Peninsula was affected by a stupendous outburst of volcanic energy, resulting in the eruption of a thick series of Lava and associated pyroclastic materials. This series of eruptions proceeded from fissures and cracks in the surface of the earth from which highly liquid lavas welled out intermittently, till a thickness of some thousands of metres of horizontally bedded sheets of basalts had resulted, obliterating all the previously existing topography of the country and converting it into an immense volcanic plateau' (Wadia, D.N., *Geology of India*, 1976: 275).

4

Historical Geography I: The Antecedents

Historical geography is an attempt at reliving the past and reinventing it. It does not study the evolution of phenomena, unlike history, which analyses the change along the time-scale. Geography recognizes time as constant and space as variable. This is an exercise in the study of a selective cross-section. An attempt has been made to reconstruct the geography of the subcontinent at different points of time in history. The first part deals with the immediate antecedents of changes in space. The early Aryan tribes migrated from the *Saptasindhava* region to *Madhyadesa* encompassing the middle and the lower Ganga plain.

Perceptions of the foreign travellers who came to India have been compared with the information contained in indigenous sources like the Vedic literature: *Brahmanas*, *Jatak* tales and the folklore. States, kingdoms, principalities, as mentioned in these texts, as far as possible, have also been located in space. In this chapter an attempt has been made to evaluate the multiple effect of the changing situation and antecedents in the context of which one should view the happenings on the political front.

India has attracted people from times immemorial. There were several phases in Indian history when the country was known as a golden bird. In fact, it will not be wrong to say that all expeditions were directed for discovering India (Landstrom 1964: 9). The author has narrated a 3,000 years history of exploration. This tale of exploration by land and by sea as told by the author covers all voyages to the Land of Punt by the Egyptians to the circumnavigation of Africa by the Portuguese thousands of years later. 'The routes were many, but the goal was the same—to reach India, or more precisely, what medieval Europeans called India'. People came from all parts of the old world, especially from East Africa, West Asia and Western Europe in conquest for India. Even Columbus, who was the leader of an expedition sent out to discover India, crossed the Atlantic Ocean and reached a group of islands, which he thought was India. The discovery of the North American continent was believed to be the discovery of India. It was in this context that this group of islands in the Gulf of Mexico was then described as Indies and the people as Indians. In the later years, India was visited by Vasco da Gama, a Portuguese mariner, who came by a sea route and landed on the west coast of India near Calicut. Later, European explorers discovered an archipelago of islands from Sumatra to the eastern most point of the Arafura Sea. These islands were also considered as India. However, in order to differentiate it from the West Indies, they were named as East Indies. Perhaps, the name of Indonesia was a mix of India and Asia. In the ancient period of history of India trading communities from South India were great seafarers and they left imprints of the Indian culture and tradition on the Southeast Asian region.

In the following paragraphs an attempt has been made to assemble the evidence on the visions of India, both outsiders and the natives. How did the people, natives as well as outsiders, describe and report their own ideas of the shape, geographical size and location of places?

Visions of India

The visions of India as recorded in Indian epics, such as the *Ramayana*, the *Mahabharata* and other literature like the *Upanishads* and the *Puranas*, have been collected and interpreted. The idea is to throw some light on the shape, extent and boundaries of the subcontinent at different points of time. These visions of the ancient Indians are characterised by the diversity of opinions and views. There are serious problems in the identification of places and the people associated with them. There are several other sources, such as the accounts of the foreign travellers who visited India at different points of time. Greek records of Alexander's campaign in northwestern India provide a storehouse of information, which is valuable in any exercise to reconstruct the historical geographies. Even the death of Alexander, while he was on his way back to Macedonia, did not result in any break in the collection of information. The Greeks continued to rule over the northwestern parts of the subcontinent for at least two hundred years after Alexander's death. Earlier, these regions were in the possession of Iranians. The Achaemenid Empire extended over large parts of the subcontinent in the north and the Iranian rule, before the Greek occupation, lasted for no less than two hundred years. There were several Chinese travellers who came to the subcontinent to visit places of Buddhist interest.

There is, of course, difference of opinion between the indigenous observers and the foreign travellers and pilgrims. The sub-altern, or indigenous sources contain a treasure of information on the shape, extent and boundaries of the subcontinent. These records can be systematically used to reconstruct the territorial organization at different points of time.

Indian epics such as the *Mahabharata*, the *Ramayana* and the *Puranas* reveal that India was an equilateral triangle divided into four smaller equal triangles (Box 4.1).

Box 4.1: Visions of India

From the accounts of the Greeks it would appear that the ancient Indians had a very accurate knowledge of the true shape and size of their country. Strabo, benefited from the accounts of Greek writers, such as Patrokles, who had visited India and collected information on India and the eastern provinces. Both Strabo and Eratosthenes, well-known Greek writers, were impressed by the accuracy. Megasthenes had actually visited Palibothra (Patliputra). On the authority of their writings, we can say that there was a general agreement on the shape of the subcontinent as a rhomboid. At a somewhat later date the shape of the subcontinent is described in the *Mahabharata* as an equilateral triangle further divided into four smaller equal triangles. Another description of India is that it was divided into nine divisions, *Nava-khanda*. It was first described by the astronomers, Prasara and Varaha-Mihira. The division of India into five great provinces would appear to have been the most popular one during the early centuries of the Christian Era... it was adopted by the Chinese pilgrims in their travel accounts (Cunningham 1975: 1).

Greek and Chinese Accounts

The early Indian texts, such as the *Mahabharata* (c. 100 B.C.) and the *Puranas* are a valuable source of information. Indian contribution to this body of knowledge was monumental. Indian astronomers, Bhaskaracharya and Varaha-Mihira (c. 560 A.C.) described the extent and boundaries of the subcontinent as they envisioned it. Several cardinal points were identified. According to them, the country extended from Kashmir in the north to Kanyakumari (Cape Comorin) in the south and from the mouths of the Indus in the west to the mouths of the Ganga in the east. Eminent Greek observers, such as Eratosthenes who visited India during Alexander's campaign (c. 326 B.C.) left behind records of their impressions of the subcontinent. The extent and boundaries of the subcontinent were expressed in linear distances between each of the cardinal points.[1]

On the basis of this record, we may reconstruct the dimensions. From Kashmir to the mouths of the Indus, the estimated distance was 3,000 stadia (345 British miles). The distance between the mouths of the Indus and Kanyakumari was estimated at 13,000 *li* (2,210 British miles). The estimated distance from Kashmir to the mouths of the Ganga[2] was 15,000 *li* (2,550 British miles); from the mouths of the Ganga to Kanyakumari, the distance was 15,000 *li* (2,550 British miles). Eratosthenes visualised India as a rhomboid.[3] According to him the Indus delimits the external boundaries of India on the west, Himalayan mountains on the north; and the ocean delimits it on the south. The shortest side was on the west. Greek writer Patrokles, estimated its length at 12,000 stadia (1,379 British miles) as against Eratosthenes who estimated the length at 13,000 stadia (1,494 British miles). From Alexander's bridge to the sea, the distance is 1,000 stadia (or 115 British miles). Then there are other estimates of distances: such as Indus to Palibothra (now Patna): 10,000 stadia (1,149 British miles); Palibothra to the mouths of the Ganga:

6,000 stadia (690 British miles); from the Indus to the mouths of the Ganga: 16,000 stadia (1,839 British miles); from the mouths of the Ganga to the Cape Comorin, along the east coast, the distance was 16,000 stadia (1,839 British miles).

'The close agreement of these dimensions, given by Alexander's informants, with the actual size of the country is remarkable and shows that the Indians even at that early date in their history had a very accurate knowledge of the form and the extent of their native land' (Cunningham 1975: 2-3). On the authority of Megasthenes, Greek geographer Strabo recorded the distance from Indus to Palibothra as 1,143 British miles; from the mouths of the Ganga to the Kanya Kumari (Cape Comorin), the distance on the map was 1,866 British miles. It may be recalled that the Indian epic *Mahabharata*, described the shape of India as an equilateral triangle. It was generally known that the subcontinent was divided into four smaller triangles. The apex of the triangle lies in Cape Comorin, the base is formed by the line of the Himalayan mountains, the base of the other triangle lies along the imaginary line connecting Dwarka in Gujarat with Ganjam in Orissa.

Indian astronomers, Parasara and Varaha-Mihira refer to *Nava-khanda* or nine divisions. According to them, Panchala was the central division, Magadha lay in the east, Kalinga in the southeast, Avanta in the south, Anarta in the southwest, Sindhu-Sauvira in the west, Harahaura in the northwest, Madra in the north and Kulinda in the northeast. It is noteworthy that these nine divisions (or *khandas*) were well known to the authors of the *Puranas*. According to Bhaskaracharaya, these divisions were named differently. They are named as Indira, Kausrumat, Tamraparana, Gabhastimat, Kumarika, Naga, Saumya, Varuna, and Gandharva.[4]

Chinese records of the second century B.C., during the reign of Emperor Wuti of the later Han dynasty, show that India was described as Yuan-tu, or Yin-tu, *i.e.*, Hin-tu and Shin-tu or Sindhu. Much later, the Arabs adopted this scheme and described the two regions as Al-Sindh and Al-Hind. The territory west of the Indus was described as Al-Sindh and to the east of the river as Al-Hind. Over centuries, changes came in our perception and nine divisions were reduced to five divisions. In fact, the concept of 'five Indies' lasted longer than the concept of nine divisions. The records of the Chinese Tang dynasty of the seventh century A.C. show that India was generally recognised as consisting of five divisions—east, west, north, south and centre. The Chinese pilgrim, Huen Tsaung who visited India in the seventh century A.C. confirmed these five divisions. Each division was recognised as a kingdom, irrespective of its size. With the changing times these kingdoms expanded or contracted depending on the geographical location, the size of the territory and the available natural resources.

In ancient India, the tradition of writing a genre of poetry, a form of royal panegyrics, was common. These were composed in praise of the legendary *Digvijayas*. The poets often described the bounds of the Bharatavarsha in hyperbolic language and in exaggerated terms. An anonymous sixth century ruler of western India was informed by his bard about the extent and boundaries of his empire.[5] His narrative is quoted below:

From the lands where the Brahmaputra flows,
From the flanks of the southern hills, thick with groves of palms,
From the snowy mountains whose peak the Ganga clasps, and From the ocean of the west,
Come vassals, bowing at his feet,
Their pride brought low by his mighty arm,
And his palace court is a—glitter,
With the bright jewels of their turbans

Source Materials

All historical geographies are reconstructed on the basis of evidence contained in the contemporary works, either on geography or history of any region of India or the entire subcontinent. We have tried to depend on substantive contemporary works of the ancient period in various exercises in the reconstruction of historical geographies. A brief overview is presented in the following paragraphs.[6]

Classical geographies of India, such as those of the Greeks in the ancient period and of the Arabs in the medieval period, are valuable contemporary source books for the historical geography of the time. The Greek accounts of India—particularly those of Diodorus, Herodotus, Megasthenes, Arrian, Strabo, Quintus, Siculus, Justin, Plutarch, Frontius, Nearchus, Apollonius, Pliny and Ptolemy—were extensively used in a number of studies. Other notable contributions came from Cunningham (1871), McCrindle's studies (1879, 1882, 1885, 1926) on the geography of ancient India and Majumdar's commentary in his *Classical Accounts of India* (1960). The value of McCrindle's work is enhanced by the fact that he annotated the accounts on geography and identified place names with their Sanskrit originals. McCrindle's work was edited by Majumdar (1927).

Migrations within the subcontinent resulted in the redistribution of population of the different racial and ethnic groups. Reference may, however, be made to the primeval tribes, such as the Munda and the Dravidians. Initially the Munda and the original Dravidian tribal groups were in mutual contact. The Mundas, however, migrated to the east. By around the same time the Indo-Aryan groups emerged on the scene in the northwestern part of the subcontinent. The contemporary Indo-Aryans in the subcontinent had genealogical relationship with the original Indo-Aryans. It is believed that they reached the subcontinent by c. 2,000 B.C. It may be recapitulated that the Indo-Aryan tribes consisted of people who communicated in one or the other of the Indo-European languages. Scholars generally agree that this group of people came from central and eastern Europe. More specifically their homeland was supposed to be located in Europe between the rivers Rhine and the Don. The existing information is not sufficient to say anything with confidence on the timing and the chronological sequence in which the Indo-European languages bifurcated in different branches. Likewise, there is not much evidence on the emergence of the Indo-Iranian branch/sub-family of languages. It is believed that the process of language formation was completed before 3,000 B.C. But dating these phases is difficult and may be disputable.

The earliest of the Hindu sacred text, *Rig-veda* refers to the Saptasindhava, Panchanada, Madhydesa and Praki. These were perhaps the first references to the emergence of regions. The geographical location of these regions suggests that the Indo-Aryans preferred to settle down along the banks of the rivers—Sindhu (Indus), Vitasta (Jhelum), Asikni also known Chandrabhaga (Chenab). References are also found to Airavati (Ravi), Parushuni or Vipasa (Beas), Satadru (Satluj) and the Saraswati. There is a general agreement that the region known as Panchanada included the plains of the five great eastern tributaries of the Sindhu river. Further east, the land between the Saraswati and Drishadvati was known to the Indo-Aryans as Brahmavarta. On the periphery of these plains lie the alluvial basins of the Gomti (Gomal), Krumu (Kurrum) and Kubha (Kabul). It is believed that these were the earliest Indo-Aryan settlements. Literary evidence suggests eastward march of the Indo-Aryans led to their settlements to the Madhyadesa regions.

Indo-Aryans in Transit

The people who established themselves as the first settlers of the Punjab, more particularly in the inter-fluvial land drained by the Indus and its tributaries, are described as the Aryans (Indo-Aryans). Archaeologists have dug out evidence of their settlements on the banks of the legendary Saraswati (now known as Ghaggar in Rajasthan, Hakra in southern Punjab and the Eastern Nara in Sind). The geographical pattern of their distribution has been inferred from the hymns of the *Rig-Veda*, the first sacred Hindu text. Their colonies emerged in what later came to be known as Saptasindhava region—the land between the Saraswati and the Kubha (or Kabul river). Comparative studies of the archaeologists and the historians have confirmed the similarities in the language and the geographical content between the *Rig-Veda* and the *Zevend Awastha*. Sapta-Sindhava has been broadly identified as the region lying in the Indus basin including its tributaries, Jhelum, Chenab, Ravi, Beas and Satluj. This was the first camp in the eastward march of the Aryan tribes. Besides Saptasindhava, their territory included Panchanada, Madhyadesa and Praki. While Saptasindhava is the land watered by the middle stretch of the Sindhu river, its tributaries and the Saraswati, Panchanada was the land of five streams, which flow in a combined channel. Similarly, the land between Saraswati and Drishadwati was known as Brahamavarta. Earlier, the Aryan tribes had their settlements on the banks of the Gomal, Kurrum and the Kubha (Kabul) rivers in the northwestern part of the subcontinent. Madhyadesa was located in the Vinasena, Yamuna and its right bank tributaries, such as Chambal and Sindh, located in the region east of the Aravalli range in Madhya Pradesh and adjoining Rajasthan states. The Saryu (Sarju) and the Rapti delimited the region in the east. Praki (Prayaga) encompassed the middle and the lower Ganga basin along with the tributaries, *viz.*, Ghaghara, Gandaki, and the Sone (Son).

Indo-Aryans on the Move

The Indo-Aryans did not stop on the right bank of the Rapti. They pushed on eastwards. The boundaries of the Madhyadesa were extended to the Rajmahal hills located on the Bihar-Bengal border. Beyond the Rajmahal hills lies the head of the Ganga delta. It appears that in their eastward march, their iron-smelting technique helped them in making strategically important tools, like the axe, iron plough and the sword. It is also believed that they were moving on their chariots driven by horses and therefore covered long distances in a short time.

Scholars, both Indian and European have added to our knowledge of the developments in the eastern regions. Religious and quasi-religious texts such as the *Vedas, Brahmanas, Sutras, Upanishads,* and the *Puranas* contain much useful material on the geographical characteristics of the regions and places where they were composed. In a number of studies this material has been organised to enable meaningful reconstruction of the geography of various regions of India at different points of time. Notable among these studies are the contributions of Bird (1840), Wilford (1822), Cunningham (1871), Law (1955), Dey (1927), Raychaudhuri (1960), Sircar (1960), Dikshitar (1952) and Ali (1966). Law based his *Geography of Early Buddhism* on the fragmentary evidence lying buried in the Buddhist texts. Dey prepared a detailed index of all geographical references occurring in the ancient Sanskrit and Pali texts. Dikshitar prepared a similar index for the *Puranas* to give some meaning to their 'vague obscurity'. Raychaudhuri and Sircar also utilised Puranic sources to sift and arrange information on the extent, topography and population of Bharatavarsha. Ali's monumental work on the *Geography of the Puranas* broke new ground. He applied the tools of modern geography to understand the real significance of geographical references in the ancient texts.[7] However, his pre-mature death in 1966 denied him the opportunity of discussing his knowledge of the period with the contemporary scholars. Fleat (1899), Abbott (1899), Clark (1919) and Majumdar (1919-1921) also drew heavily from these sources. More specifically, scholars like Stein (1971), Das (1921) and Bhargava (1956; 1964), used geographical material in the *Rig-Veda* and Tamaskar (1956; 1967-68) and Mukerji (1969) used similar material from the *Upanishads* and the *Jataka* tales. Dubey (1967) deserves special mention since he has utilised a variety of source materials on ancient India.

As is generally known, the kingdom of Kosala with its capital located at Ayodhya, was the scene of the epic *Ramayana*. The great epic is rich in geographical content, which has been utilised, among others, by Das (1894), Pargiter (1894) and Dikshitar (1935). Similarly, the geographical material in the *Mahabharata* has been painstakingly used in all major contributions to the historical geography of ancient India. Literary sources, such as the works of Kalidasa, were helpful in the understanding of the geogrpahy of the period by Sivaramamurthy (1932), Law (1954) and Gupta (1963). New Janapadas emerged in north Bihar and the adjoining parts of Uttar Pradesh, like Mithila, Kasi and Magadha. By the time of *Mahabharata* we come to know about the Kauravas and the Pandavas and the war fought between them. We also hear about the Kuru and Panchala kingdoms.

Supposedly the southward thrust of the Indo-Aryan speaking tribes resulted in the establishment of other nodes of power, such as Matsaya and Surasena lying in the region to the west of the Yamuna and adjoining Rajasthan. The Indo-Aryans moved southwards and settled on the banks of Godavari and the Tapi rivers. We also know about other kingdoms like Kalinga, in what is now known as Orissa, and the Sindhu and the Salva kingdoms in Sind and Rajputana respectively.

Travel accounts constitute an invaluable source of information for the historical geographer. These records offer the rare opportunity of having an eyewitness account of the lands and the people visited by the travellers at a given point of time. In the Indian context, most valuable among these travel records are the narratives of the Chinese pilgrims, like Huen Tsaung, and to a lesser degree, Fa-Hain and Sung-Yun. The enormous evidence in Huen Tsaung's, *Si-Yuki* was ably sifted, annotated and analysed by Beal (1957-58), who took great pains in identifying place-names and other geographical material mentioned by Huen Tsaung. This evidence was utilised by Cunningham (1871), who not only cast it in the geographical mould but also compared it with the accounts of Alexander's campaign to produce a brilliant geographical account of Buddhist India.

Encounters with the West

By around 500 B.C., a vast chunk of the territory of the subcontinent was ceded to the Achaemenid's Empire. The Persians (Parthians) controlled the territory up to the Indus in the northwest. In fact, the Indus river was the eastern limit of the Persian Empire. The Empire extended in the north and the northwest from Indus to the Kabul river and the regions beyond the Hindu-Koh Range in Afghanistan.

Alexander, the prince of Macedonia, displaced the Iranian King Emperor Darius, and annexed his empire. Alexander's eastward thrust proved to be a deathblow to the vast Persian Empire. The Greek prince was challenged by Porus, who was the king of the territory between the two rivers—Hydaspes (Jhelum) and Akesines (Chenab). Despite his military power consisting of foot soldiers and elephants, Porus was defeated and surrendered. The Greek prince found him as an impressive personality, well built body, tall and handsome. Alexander looked at him and decided not to execute him. As a concession to him, Porus was allowed to retain his kingdom but as a vassal of the Greek Emperor.

One of Alexander's generals, Seleukus Nikator, who stayed back after Alexander's return, controlled all land from Indus to the Hindu- Koh. Chandragupta Maurya, the king of Magadha challenged and defeated the Greek army in 305 B.C., seventeen years after Alexander's invasion, with the result that several provinces, such as Paropamisedal, Aria, Arachosia and Gedrosia were ceded to the Mauryan Empire.

Serial Invasions

Wanderers like the Sakas (Indo-Scythians) intruded into the Sindhu river basin and the adjoining parts of Saurashtra. The Saka chief adopted the Persian title of Satrap. In the next stream of intruders came the Yavanas, who are generally considered as Asiatic Greeks or Greco-Bactrians, and the Pallavas also known as

Parthians held their sway on northwestern India. But their rule did not last long. The picture that emerges in the beginning of the second century A.C. is that Andhra kings controlled the territory in western India. But a little later several parts of the territory, particularly Saurashtra, Malwa, Kachchh, Sind, and the Konkana region enjoyed an independent status under the powerful king Rudraman, himself a Satrap. Sogdiana, Bactria, Kabul valley and the Punjab plains were controlled by the Greeks and the Parthians. However, they were overthrown by Yuch-chi, hordes of Turkik nomads who in course of time settled down in this area. Kanishka, the third king of the Kushana dynasty emerged as a powerful king. Kanishka's empire extended up to Persia. It also encompassed Kashgar, Yarkand and Khotan, important cities in the Central Asian region. The empire also extended over the northwestern region of the subcontinent. The Jumna and the Chambal rivers were supposed to be the eastern limits of the empire. Purushapura, or Peshawar, was the capital city of Kaniska's empire. It is noteworthy that the emperor controlled such a vast territory from a peripheral node like Peshawar.

Foreign Elements in Recent Times

Historical records show that the subcontinent has always accommodated people from outside. There are China towns in Calcutta and other metropolitan cities of India. The latest example is of the Tibetans who came along with the Dalai Lama after Tibet was annexed by China. The Tibetan influx into India started in March 1959 when His Holiness the Dalai Lama fled to India seeking refuge. The Government of India granted him political asylum. There were many Tibetans who came along with His Holiness and many others followed him. By 1961, the Tibetan population in India numbered 43,000. The population increased to about 50,000 in 1971. A recent census conducted by the Central Tibetan Administration based in Dharamsala enumerated 97,000 Tibetans living in different parts of India (Ahmad 1995: 7-14).

The Indian Tibetans live in settlements, or in camps in the urban metropolis. Land is allotted by the state governments. Besides, there are exclusive Tibetan educational centres with boarding facilities. A substantive population of Tibetans has been living on a self-supporting basis in different parts of the country. Politically all of them have a refugee status. To stay in India, each Tibetan adult needs a registration certificate, which is issued by the government of India. It is on the strength of this certificate that he/she can travel abroad. The Tibetans do not enjoy any of the Indian citizenship rights. In the initial stage, almost all Tibetans were concentrated in the sub-Himalayan districts of the states of West Bengal and Uttar Pradesh. Later it was thought necessary to shift them to other states. They were settled in Tibetan settlements, exclusive camps and colonies. Today, they are living over a vast region in north India as well as the states of the peninsular south. The Darjeeling district of West Bengal continues to be a significant node.

Buddhism is the basis of their cultural identity, although their population subsumes some other elements of other religions. While all the Buddhists belong to the Mahayana branch, there are at least five different schools of Buddhism within India. These schools are Kagyu, Sakya, Nyingma, Geluk and Bon. Each school has a spiritual teacher who is the head of the sect. There are differences in the practice of Buddhism. In fact, they represent different paths to *nirvana*. *Nirvana* is self-sustaining; it means that one has to die before the self, in order to accommodate others to live in harmony. Harmony with nature and co-existence are the cardinal principles of the faith, irrespective of the variation of sects.

The Tibetan settlements in India are distributed in the northern, northeastern and the southern states of the country. These settlements accommodate about one hundred thousand Tibetans. Their geographical distribution shows that they are mainly concentrated in Himachal Pradesh, Jammu and Kashmir (Ladakh), Delhi, East Uttaranchal, Chhattisgarh, Jharkhand, Orissa, Sikkim, Arunachal Pradesh and Meghalaya.

Strangers in their Own Land

The south Asian nations have often shown concern on the influx of refugees. India, for example, received a vast chunk of displaced persons belonging to different communities like Hindus and Sikhs from Pakistan and Bengali Hindus from East Pakistan (now Bangladesh). Similarly, Pakistan received the so-called Muhajirs (refugees) from Uttar Pradesh, East Punjab, Bihar, Madhya Pradesh, Rajasthan and Gujarat. Sri Lanka received Tamils who moved to the island at different points of time. Bangladesh had its share of refugees in Bihari Musalmans who were lawfully citizens of Pakistan and therefore enjoyed citizenship status in Pakistan even after Bangladesh became free in 1971. Biharis are trapped in makeshift camps ever since the inception of Bangladesh as an independent country. The paradox is that they were the same people who launched the movement for Pakistan.

There were residual problems after the partition of India in 1947, which have been inherited by India, Pakistan and Bangladesh. People have fled away from the Jammu and Kashmir state for reasons of their security in the wake of growing civil strife in the valley of Kashmir. All accounts show that these displaced persons—both Kashmiri Pandits and Kashmiri Musalmans—came out of their state under pressure. The Kashmiri Pandit families live in miserable conditions in camps outside Jammu. Kashmiri Musalman families have also preferred to leave the state under the fear of terrorist attack and extortion. The Bengali Hindus who migrated from Bangladesh were allotted land in parts of Madhya Pradesh, Chhattisgarh and Orissa. They live in the midst of *Janjatis* who had little idea of irrigation based intensive agriculture. The Dandakaranya Development Authority had assigned them land in Bastar (Chhattisgarh) and Koraput (Orissa) districts. The history of Development Authority is full of episodes when the Bengali immigrants alienated the tribes from their traditional land.

The Tamils in Sri Lanka are a different case altogether. They are the descendents of the first generation Tamil immigrants who around 1840 migrated to Sri Lanka as hired labourers. The Tamils in Sri Lanka are mainly settled in the Jaffna Peninsula. They are also living in the interior hilly region. A second wave of migration consisted of agricultural labourers, as there was scope for them to find employment on plantation estates. It is believed that the Tamil problem got aggravated when they insisted on a dual citizenship. They also demanded that Tamil language should be recognised as one of the principal languages of the island. This became the main source of tension between the Tamils and the Sinhalese. India participated in negotiations with Sri Lanka, although with little success. At one stage the two negotiating sides offered for 525,000 Sri Lankan Tamil's return to India. There have been periodic migrations and return migrations—both continuing over the period up to 1950. The 1961 census recorded a total population of 28,600 Sri Lankan Tamils rehabilitated in different parts of the states of Tamil Nadu, Karnataka, Andhra Pradesh and Maharashtra. It was expected that both the Sinhalese and the Tamils would sort out their differences and take necessary steps for reconciliation. The contribution of Tamils to the development of Sri Lanka cannot be overstated.

A Resume

The subcontinent was always an object of curiosity. In fact, it is a mystery that the region attracted foreigners from times immemorial. The history of the exploration of the world was really a quest of India. The maritime powers of Europe who launched exploratory exercises always tried to discover India. Paradoxically, when they reached the unknown parts of the world, they named them Indies or India. Examining the history of India, one may note the duality of the north and the south. The argument is that north and south were two different macro-regions. Expansion and contraction of the empires/states, kingdoms, and chieftainships, can be studied in the context of north-south duality. There were phases when the northern region was divided into territories. These phases were followed by a vigorous process of unification. Evidently, unification and disintegration chased each other in a mysterious way. The cross-sectional approach leaves scope for reinventing the past. It may be a disputable issue whether one can reconstruct the past as a whole. But reconstruction is a necessity and the success of the research project depends upon the database. It has been noted that the northern parts of the subcontinent from Afghanistan to Assam were more exposed to external influences. But all that happened in the north left its imprints on the south. Many-a-time north-south duality has been exaggerated. But it would be difficult to say that south was impregnable at any point of time. The south always remained exposed to the forces centred in the north and *vice-versa*. This is evident from the fact that the Mughal Empire (1526—1707) remained by and large confined to the region north of the Narmada trough and the Vindhyan highlands and mountain complexes from west coast to Chhotanagpur plateau in Bihar. The conflict between the Vijayanagar and the Bahmani kingdoms was an example of the encroachment of the north on the south. The Andhra kings at one point of time ruled over remnants of the Maurayan Empire in Magadh and Kalinga kingdoms. Even today, the same laws of land apply and the polity of the subcontinent is a result of the balancing of centrifugal and centripetal forces. It may be concluded that the Indian union today is not the end of the road. At any stage of time, the centrifugal forces may surpass the centripetal forces. The political situation is fluid and is likely to remain fluid in the future also.

END NOTES AND REFERENCES

1. According to the Mahabharata (c. 100 B.C.), cardinal points were as follows: Kabul river in the northwest, Brahmaputra in the northeast, Ganjam in the east, Dwarka in the west and the Cape Comorin in the south.
2. The delta of the Ganga and the Brahmaputra rivers in Bangladesh.
3. An unequal quadrilateral.
4. Bhaskaracharya and Varaha-Mihira differed on the names of the nine divisions.
5. Translation from an inscription of Yashodharan, of Mandsaur, cit. Basham, 1964: 6.
6. Aijazuddin Ahmad (1972): 'Historical Geography' in *A Survey of Research in Geography*, Popular Prakashan, New Delhi, pp. 147-169.
7. The reference is to the reign of Empire Wuti of the later Hun dynasty of China in the seventh century A.C. Indo-Aryans, the last great race movement into India was associated with the Vedic invaders who belonged undoubtedly to the Northern Steppe folk: moving southwestwards in a great racial wave along with the Kassites, they swept into northwest India somewhere during the second millennium B.C. It is possible that this type was first characterised in the Euro-Asiatic steppelands, between southeastern Russia and southwestern Siberia, from some of the more powerfully built long-headed races who drifted towards the northeast with the retreat of the ice sheets. In the *Kurgans* (burial mounds) of southern Russia and in the Minusinsk region of the Upper Yenisie of the early Bronze Age, skeletons conforming to this type have been found, together with evidence of horse-sacrifice. Chinese chronicles also speak of a tribe called the Yuechi who established the Tokharian state in eastern Turkistan in the centuries preceding the Christian era. They were described as light-skinned and red-haired, and as Tokharian has been found to belong to the *centum* group of the Indo-Aryan languages, closely related to Greek and Latin, the Tokharis or Yuechis can reasonably be supposed to have been Indo-European-speaking people (Guha, B.S., 1944: *Racial Elements in the Population*, Oxford Pamphlets on Indian Affairs, Oxford University Press, p. 23).

5

Historical Geography II: A Tale of Five Empires

This chapter deals with the information in contemporary sources to show that the state evolved from primary nodes to the empires. Facts, which are antecedental to this development, have been collected and analysed. The process of territorialisation has been analysed in the context of time and space. The development was related to the incoming of the Indo-Aryans and their expansion within the subcontinent. The story of the past is narrated in the light of migration of the Indo-Aryans from the upper plain of the Indus to the deltaic region of Bengal. The cut-off points are fourth century B.C., seventh century A.C., fourteenth century A.C. and the beginning of the eighteenth century A.C.

The Mauryan Empire (c. 232 B.C.)

The first Indian empire rose from the kingdom of Magadha, south of the Ganga in Bihar and the nerve centre of the Mauryan Empire. The extent and boundaries of the empire show how powerful the Magadhans were. It was the first ever-Indian empire. The empire was dependent on the agricultural surplus and the trade. The founder of the empire, Chandragupta Maurya, ruled for twenty-four years. His dominions extended from the Bay of Bengal to the Arabian Sea encompassing the entire mid-Indian region. The Mauryans were also controlling the territory from the Himalayas to the Hindu-Koh.

The Mauryan Empire incorporated the entire north Indian plain from Rajmahal Hills to the Hindu-Koh Range. It consisted of most of Afghanistan, Baluchistan and Sind. It may be noted that the empire also controlled other kingdoms, such as Kashmir, which were also under the possession of Ashoka, grandson of Chandragupta Maurya. He built the city of Srinagar in Kashmir and Lalitapatan in Nepal. His authority extended over Bengal, Kalinga and several other kingdoms in the east. Parts of Deccan were also in the possession of the Mauryan kings, particularly Ashoka. He was accepted as a suzerain power over most of India. The peninsular region between the Godavari and Krishna rivers was, however, in the possession of local kings. But they accepted the Mauryan king emperor as the authority and were forced to pay taxes imposed by the king. Further south lay the Tamil kingdoms of Chola, Pandya, Satiya and Chera. New centres of power, like Andhra, emerged and extended their hegemony over the territory up to the sources of Godavari all along the peninsula from the Bay of Bengal to the Arabian Sea. Perhaps, as some historians believe, the tables turned. Magadha itself came under the suzerain authority of the Andhra kings and it seems that the situation remained the same for many years.

Harsha Vardhana's Empire (c. 606 A.C.)

We have considered the first half of the seventh century A.C. as a landmark in the long history of the subcontinent. The entire region came under the authority of king Harsha Vardhana, the king of Kannauj. A Chinese pilgrim, Huen Tsaung has left behind an account of Harsha's territories throwing light on contemporary political organisation, culture and the people.

Pages from the Diary of a Chinese Pilgrim

The Chinese pilgrim came to India in early seventh century A.C. He spent around fifteen years visiting important places of Buddhist interest in the subcontinent. His travel account was translated into English by European Indologists. He recorded the broad features of the political organisation and social life. Based on his work, an attempt has been made here to interpret the information on the contemporary geography of India.

The regional/political structure of the subcontinent, as reported by Huen Tsaung, has been reconstructed in the following paragraphs. According to the traveller, the subcontinent was divided into the following kingdoms:

The northern region consisted of parts of eastern Afghanistan, Kashmir, Punjab and the adjoining hill states in what is now known as Himachal Pradesh; beyond the Indus and all regions between the Saraswati and the Indus rivers. The western region included Sind, western Rajputana, Kachchh, Kathiawar and the traditional Gujarat, adjoining coastal region, along with the lower basin of the Narmada river. The central region comprised Ganga valley from the Indo-Ganga-divide region from Thanesar to the Rajmahal hills and the head of the Ganga delta, and from the Himalayan mountains in the north to the Narmada river in the south. The eastern region extended from Assam, Bengal, delta of the Ganga together with Sambalpur, Orissa or Odra to Ganjam on the east coast of Orissa. The southern region consisted of the Peninsula from Nasik and Ganjam to Cape Comorin, encompassing Berar, Telengana, Maharashtra and Konkana. It also included Hyderabad, Mysore and Travancore, the whole of the peninsula, south of the Narmada and the Mahanadi rivers (Table 5.1).

Table 5.1: Kingdoms Visited by Huen Tsaung (A.C. 629—645)

North

I. Kaofu or Afghanistan

1. Kapisene or Opian
2. Kophene or Kabul
3. Arachosia or Ghazni
4. Lamghan
5. Nagarahara or Jalalabad
6. Gandhara or Parashawar
7. Udyana or Swat
8. Bolor or Balti
9. Falana or Banu
10. Opokien or Afghanistan proper

II. Kashmir

1. Kashmir
2. Urasa
3. Taxila or Takshasila
4. Sinhgapura or Ketas
5. Punacha or Punch
6. Rajapura or Rajori

Hill States of the Punjab

Jalandhara, Champa or Chamba, Kullu, Mandi, Sukhet, Nurpur/ Pathaniya, and Satadru (Sutlej)

III. Taki or Punjab

1. Taki or northern Punjab
2. Shorkot or middle Punjab
3. Multan or southern Punjab

West

I. Sindh

1. Upper Sindh
2. Middle Sindh
3. Lower Sindh or Lar
4. Kachchh

II. Gurjjara or Gujarat

III. Valabhadra or Balabhi

1. Balabhi
2. Saurashtra
3. Bharaoch or Barygaza

Centre (Madhya Desha)

1. Sthaneswara or Thanesar
2. Bairat
3. Srughna
4. Madawaar
5. Brahmapura
6. Govisana or Kashipur
7. Ahichhitra
8. Pilosana
9. Sankisa
10. Mathura
11. Kanoj
12. Ayuto or Ayodhya
13. Hayamukha
14. Praya
15. Kosambi
16. Kosapura
17. Visakha, Saketa
18. Saravasti
19. Kapila
20. Kusinagara
21. Varanasi or Banaras
22. Garjjapatipura
23. Vaishali
24. Vriji
25. Nepala
26. Magadha
27. Hiranya Parvat
28. Champa
29. Kankjol
30. Paundra Vardhana
31. Jajhoti
32. Maheshwarapuram
33. Ujjain
34. Malwa
35. Kheda
36. Anandapura
37. Vadari or Eder

East

1. Kamarupa
2. Samatata
3. Tamralipti
4. Kirana-Suvarrna
5. Odra or Orissa
6. Ganjam

South

1. Kalinga
2. Kosala
3. Andhra
4. Donakakotta
5. Choliya or Jorya
6. Dravida
7. Malakuta or Madura
8. Konkana
9. Maharashtra

Ceylon

Source: Compiled by the author on the basis of information from Cunningham 1975; Ali 1966.

Historical records show that the Chinese traveller reached the subcontinent in May 630 A.C. He stayed and wandered in Afghanistan for some time. We are told that the pilgrim reached *Al Hind* after crossing the Indus river in April 631 A.C. He visited Taxila and spent several months in familiarising himself with the holy texts of Buddhism. He reached Kashmir by the Jhelum gorge route. His stay in Kashmir for about two years proved to be very useful for his mission. He discovered several records and acquired the working knowledge of language, the Sanskrit. Evidence shows that the famous Sangla fort was lying in ruins. Sangla was one of the places, which came into limelight during Alexander's invasion of the subcontinent. The Chinese traveller stayed on for about fourteen months at a place called Chinapati. He also stayed in Jalandhara for about four months. He reached the Satluj in 635 A.C. His easterly route was, however, circuitous. This is understandable because every time he visited places of his interest, he discovered that there were other places, which he had left behind. He reached Mathura and then decided to go to the north. The northern route took him to Thanesar, located about 200 miles (320 km.) from Mathura. From Thanesar he took the route via Srughna on the Jumna and via Gangaduara (now known as Haridwar). He reached Ahichhatra, the capital city of the northern Panchala kingdom. He visited Sankisa, Kannauj and Kosambi and crossed the Ganga once again to visit Ayodhya and Saravasti. He then visited Kapilavastu and Kusinagara—places associated with the life of Buddha. He also visited the city of Banaras and Sarnath. Moving in an easterly direction, he came to Vaishali from where he made a short visit to Nepal.

He resumed his journey eastwards and reached Patliputra (also spelt as Palibothra). His next destination was Bodh Gaya. He visited Kusagarapura and Rajagriha (Rajgir), capitals of the kingdom of Magadha. He spent fifteen months at the famous monastery of Nalanda. From Nalanda he followed the route lying parallel to the Ganga to reach Modagiri and Champa (Champaner). He also visited Paundra Varddhana (Pubna) and Kamrupa (Assam). From here he took a route to south India. During his southward travel he passed by Samatata (Jessore), Tamralipti (Tamluk), Odra (Orissa) and reached Ganjam and Kalinga. He visited Kosala (Berar) early in 639 A.C. He passed through Andhra (Telengana) and Dhanakakata (Amravati). Early in 640 A.C. he reached Kanchipura (Conjeveram), the capital of the Dravida kingdom. From Dravida he took the northern route and passed through Konkana, Maharashtra and Bharuch. During this travel he visited Ujjain and Balabhi. Finally, he reached Sind and Multan on his way back home. We are told that he lost important manuscripts collected during this trip. He therefore decided to go back to Magadha once again. It appears that his purpose was to remove his doubts about Buddhism. During this revisit he contacted a noted Buddhist teacher, Prajnabhadra. Then he proceeded to Kamrup and by 643 A.C. he came to Patliputra. At Patliputra he received audience of the king Emperor Harsha Vardhana (also known as Siladitya). He travelled to Prayag, Kosambi and Kannauj in the company of king Emperor Harsha. Finally, he crossed the Indus river around 643 A.C. on his way back to China.

Huen Tsaung's itinerary gives him the credit of being a great traveller of the world. He stands in the company of celebrated explorers, like Ibn-e-Battuta and Marco Polo.

Huen Tsaung's travel account, *Si-Yuki*, throws new light on the political organisation of the subcontinent in the seventh century A.C. The Chinese traveller came to seek first hand knowledge of the Buddhist religion. Huen Tsaung was perhaps the most enterprising among the Chinese travellers. His travels in the subcontinent during the period 629-645 A.C. are of great value to the historians. He maintained a record of his travels. It is necessary to organise the heritage of the subcontinent. British scholars/ administrators along with German and French visitors to the subcontinent in the later years did remarkable work on the basis of the floating knowledge gained by a methodical study of the documents. It was realised that the identification of places, mentioned by Huen Tsaung, was difficult. Also, their location and estimates of distances between places was an uphill task. Mention may, however, be made of Alexander Cunningham's monumental contribution to this body of knowledge.[1]

Eighty Kingdoms, Seven Confederacies and One King Emperor

Huen Tsaung's account mentions eighty autonomous kingdoms, which were under the suzerain authority of King Emperor Harsha Vardhana. Each of them had a separate ruler (*raja*), and all of them were tributary to a few of the greater kingdoms. For example, northern India consisted of Kabul, Jalalabad, Peshawar, Ghazni and Banu kingdoms. All of them were tributary to the ruler of Kapisa, whose capital city is identified as Charikar or Singhapora. Kingdoms, such as Urasa, Punach (Punch) and Rajauri, were tributary to the *raja* of Kashmir. Similarly, the Punjab, Multan and Shorkot were tributary to the ruler of Taki, or Sangala, located near Lahore.

Harsha Vardhana was holding sway on the central and the eastern parts of the subcontinent from Sthaneswara (modern Thanesar) to the mouths of the Ganga, and from the Himalayan mountains to the banks of the Narmada and Mahanadi rivers. These kingdoms were tributary to the great King of Kannauj. It appears that Jalandhara was also tributary to him, probably as the evidence shows; the ruler of Taki might also have been a tributary to the kingdom of Kannauj under Harsha Vardhana (Box 5.1).

Box 5.1: Centre Vs. Region

'The Rajput king of Maharashtra ...was perhaps the only sovereign who had successfully resisted the armies of Kannauj. This statement of the Chinese pilgrim, Huen Tsaung's view is corroborated by several inscriptions of Chalukya princes of Maharashtra....' 'The great king Harshavardhana was the paramount sovereign of 36 different kingdoms, comprising nearly one-half of India in extent, including all its richest and most fertile provinces.'

'The extent of his dominions is clearly indicated by the names of the kingdoms against which he directed his latest military campaigns. These kingdoms included Kashmir in the northwest, Maharashtra in the southwest and Ganjam in the southeast. Within these boundaries the paramount ruler of the subcontinent of India was Harshavardhana'[2] (Cunninghum 1975: 10-12).

Records show that 18 tributary princes attended on their suzerain lord during his great religious procession from Patliputra to Kannauj in 643 A.C.

At the time of Huen Tsaung's visit, South India was divided into 9 kingdoms. These were Maharashtra and Kosala in the northern part of the peninsula; Kalinga, Andhra, Konkana and Dhanakakta in the middle, and Jorya, Dravida and Malakuta in the south.

Tughlaq's Empire (c. 1347 A.C.)

Delhi Sultanate: Pages from the Diary of a Moorish Traveller

The king Emperor Mohammad bin Tughlaq ruled over India in the fourteenth century A.C. Ibn-e-Batuta visited the subcontinent around 700 years after the Chinese pilgrim Huen Tsaung. The political scene had changed drastically. Early fourteenth century saw the emergence of a new political order. This is the story of the Tughlaq dynasty. Delhi Sultanate became the hub of the new religious and political superstructure. Ibn-e-Batuta's travel account, the *Rehla,* throws ample light on the social structure and the political development of India in the first half of the fourteenth century. His narrative was translated, edited and annotated by Gibb (1929; 1962). Yule (1874) arranged the geographical content on India in the *Rehla,* while Haig (1887) described his travels in Sind, identifying place-names and assessing the geographical significance of the narrative.

Ibn-e-Batuta's Itinerary

Ibn-e-Batuta was born in Tangier on February 24, 1304. He left for Mecca for the Haj pilgrimage. From Mecca he went to several places in West Asia and southeastern parts of Europe. Our interest is, however, only in his travels in the Subcontinent during the period extending from June of 1325 to April of 1347. His *Rehla* provides a storehouse of information on the places he visited and the people he met. It appears that he travelled by sea and reached the port of Lahari, near the mouths of the Indus. His route took him to Sehwan (Siwasitan), Janani, Bukkur and Uch all in the Sind province of Paksitan. He reached Multan on November 5, 1333. In his travel from Multan to Delhi he passed through Pakpattan, Abohar (Abu Bakhar), Saraswati (Sirsa), Hansi, Masudabad and Palam. He reached Delhi on June 9, 1334. He visited Delhi when the Tughlaq King, Mohammed Shah, was in power. The Emperor was impressed by his knowledge of Islamic law and jurisprudence and his qualities of head and heart and appointed him as the Qazi-ul-Qazzaat (Chief of the Qazis). Then he expressed his desire to go to other parts of India. He left Delhi for Bijnor via Amroha. He then went from Amroha to Afghanpur across the Sarju river and came back to Delhi in 1336. The traveller proceeded on his journey from Delhi to Sargadwari. From there he travelled to Kannauj and Bahraich and returned to Delhi in 1340. In 1341, he went from Delhi to Sehwan to meet the Emperor once again and returned to Delhi. In the meantime he was given an assignment to go to China as an ambassador of India.

On his way to China he travelled via Tilpat, Aou, Hiḷar, Bayana, Koil, Jalali,

Tajpur, Brijpur, Kannauj, Hanaul, Wazirpur, Jalesar, Mawri and Marh (Mahrera). He reached Marh on September 18, 1342. However, it is not clear as to which route he took to reach Alapur on his way to Gwalior. But it is reported that Ibn-e-Batuta visited Gwalior on September 25, 1342. His journey from Gwalior to Chanderi started on Septermber 27, 1342. The visitor reached Perwan on 27 September of the same year. His wanderings in central India from Amwarhi to Khajrao and Chanderi implied that he wanted to visit places, which were known for their architectural excellence and were important trading centres. He reached Chanderi on October 9, 1342 and reached Dhar on October 28, 1342. From Dhar he travelled via Ujjain to Daulatabad, an important centre south of the river Tapti (Tapi) which was selected by the Tughlaq King to be the new capital of his empire. From Daulatabad he went to Cambay via Nandurbar on the west coast of India. His voyage from Cambay started on November 27, 1342 and took him through Kawa, Gandhar, Perim, Gogo, Sandapur, Hinawar, Barcelore, Fakanar, Manjarur, Phili, Jurfattan, Dahfattan and Budfattan, ports lying on the west coast. He reached Budfattan on December 31, 1342. The voyage to China took him to Calicut, where he halted for eighty-eight days at a stretch.

His voyage along the Malabar Coast from Cambay to Quilon took him to the following ports: Sandabur (a place near Goa), Hinawar, Abusarur, Fakanur, Manjarur, Phili, Jurfattan, Dahfattan, Fandarayna and Calicut. From Calicut he proceeded to Quilon via Saliyal and Kunjakari. Then he came back to Calicut by a land route from Fattan to Quilon via Madura and from Quilon to Calicut on the sea. During his stay of eighty-eight days in Calicut, he went to the Maldive Islands. The ports he visited included Kannalus, Mahal and Muluk. From Muluk he returned to Battala on the west coast of Sri Lanka. He visited places such as Salawar, Kunakar, Dinawar, Qali, Colombo, all ports lying on the west coast of Ceylon and then he came back to Battala. From Battala he travelled to Arcot (then known as Harkatu) from where he came back to Fattan. His journey along the Malabar Coast and the Coromandel Coast can be divided into the following segments: Malabar to Maldives, Maldives to Ceylon, Ceylon to Mabar, Mabar to Malabar and from Malabar to Maldives. He had a knack for wandering.

The Mughal Empire (c. 1707 A.C.)

Much water has flown down the rivers Jumna and Ganga since the Mughals built a vast empire in South Asia. The Mughals established a dynasty of which Babar was the founder and Aurangzeb, the last of the Great Mughals. The dynasty continued after Aurangzeb's death, but only in namesake. None of his descendents were capable of holding such a vast territory. The times had changed.

Aurangzeb Alamgir (1658-1707 A.C.), the last of the Great Mughals died in 1707. The year marks the beginning of the end of the Mughal Empire. But it also marked the dismemberment of the Indian state. It was under him that a central authority exercised its power to keep the peripheral regions together. There is evidence to show that a strong centrifugal force was in operation on the periphery. Aurangzeb was, perhaps, the only Mughal king who taught the art of the statecraft to his sons. He took pains to prepare them and train them as the future rulers (*Ruqaat-e-Alamgiri*—Letters of Alamgir to his sons). But the tide turned against the Mughals.

A hundred years later the British Empire rose on the ashes of the Mughal Empire. But the British granted the local princes a certain amount of autonomy. There were situations in which the princely states were found more authoritarian than the British themselves. Moreover, the princely states, by and large, did not encourage their subjects to participate in the on-going freedom struggle. Not only that, some of the princely states 'banned' the entry of the nationalist leaders in their states; they left no stone unturned to vie with each other to keep the British in good humour. At independence in August 1947 the Indian state was at the highest degree of fragmentation. As a first step towards the integration of the country the princely states were persuaded to merge with the Indian Union. The end of the British rule came with the division of the country into two nations—India and Pakistan. The two nations theory had a popular appeal. The Muslims of India rallied behind Muslim League. Twenty-four years later, Indira Gandhi reinvented the two-nation theory and helped emerge the third nation—Bangladesh. Seen in this context, Aurangzeb's Empire was the last monument of Indian unity and territorial integrity. The Mughals in India were an integrative force, not colonialists. But they were imperialists. This is also true that Aurangzeb's expansionist policy met stiff opposition from the regional or local powers, such as the Rajputs, Jats, Sikhs and the Marathas, who were struggling for independence from the Mughal rule.

The Mughal king emperor brought under his sway a vast chunk of territory from the Hindu-Koh in the west to the mouths of the Ganga in the east and from the Himalayas to the mouths of the Kaveri in the south. The tip of the peninsula was in the possession of the Polygars, who had a large measure of autonomy in their internal matters. But they were also forced to pay tribute to the Mughal king. The only part of the subcontinent, which was free in the technical sense of the term, was the region from the mouths of the Kaveri to the southern tip of the peninsula and from Cape Comorin to the Malabar Coast.

The foreign maritime powers were interested in trade with India under Aurangzeb. In fact they were waiting for his death. Several European maritime powers, such as the Portuguese, French, British, Dutch and the Danes were waiting for the emperor's death. The foreign interest was already there in the earlier centuries, but it remained confined to the coastal region. They were allowed to establish their factories on the west coast, as well as on the Coromandel Coast by the Mughal kings before Aurangzeb. But they were not allowed to expand their territory on the mainland. The European maritime powers fought among themselves as well as in combination with the local/regional powers. The old king's death created a political vacuum, which the foreign powers were going to fill. It may be noted that Aurangzeb proved to be the main hurdle in the ascendancy of the colonial powers in the subcontinent (Box 5.2). After his death his empire was divided among the regional aspirants to power—Afghans, Rajputs, Rohilas, Nizam, Mahrattas, and Mysore. The Maratha confederacy extended from Jumna to Goa and from coast to coast in the peninsula. Hyderabad and Mysore were vast territories. In fact, the peninsular interior was all in all in the possession of Nizam of Hyderabad and the Maharajas of Mysore and Travancore.

Box 5.2: European Maritime Powers in the Subcontinent

The year 1700 is a convenient time for enumerating the European possessions so far acquired in India. The Portuguese were the first on the scene. From the landing of Vasco da Gama in 1498 near Calicut to the appearance of the Dutch and English at the end of the sixteenth and the beginning of the seventeenth century, they had a monopoly of trade between Europe and India. They took Goa in 1510, fortified Chaul in 1521, acquired Diu, Bassein, and the Bombay islands in 1534, and Daman in 1559, all of which except Bombay they still possessed in 1700.

The chief settlements of the Dutch founded in the seventeenth century were: Negapatam, Sadras, Pulikat, Bimlipatam, and Cochin. Although the English East India Company (founded in 1600) had established factories at Surat (1611), Calicut, Masulipatam, and other places, they built their first fort (St. George) at Madras only in 1639. Bombay, given in 1661 to Charles II, as part of the marriage portion of Catherine of Braganza, was finally handed over to the English East India Company in 1669. Fort St. David, south of Madras, was acquired by purchase in 1691. In 1696 the villages of Chuttanatti, Calcutta, and Govindpur were purchased from Azim-u-shan, Aurangzib's grandson. In 1700 the French possessed Masulipatam (1669), Pondicherry (1674), and Chandarnagar. At about the same time the Danes held Tranquebar and Serampur (Joppen 1917: 14).

Vijayanagar Empire (c. 1374 A.C.)

India South of the Narmada

The west-flowing river Narmada may be described the northern border of the South. Its trough along with a series of mountain ranges running from the Western Ghats to the Ranchi plateau has worked together to restrict movement on either side of the natural barrier. A continuation of the same topography is found in the east-northeast of the Narmada, particularly in the tract of land where headstreams of the Mahanadi river are located. However, the peninsula south of the Narmada is due to its relief. The southern region constitutes the Dravidian heartland. Maharashtra, mainly due to its geographical location, appears to be a transitional zone between the Dravidian south and the Indo-Aryan north. The existing geographical distribution of the Marathi speakers leads to the conclusion that the Marathi language was an enclave of the Indo-Aryan Marathi speech realm between the Dravidian speech forms—Tamil, Telugu, Kannada, and the Malayali. In between are found the pockets of Astro-Asiatic languages spoken by different tribal communities. It will be a wrong conclusion if it is argued that this zone has not permitted north-south movement of the people and the linguistic and cultural traits of the groups living in this area. While the southern kingdoms have been rising and declining, the south has always retained its ethno-lingual core intact through the ages. Civilisations have come and gone. But the dominance of the Dravidian languages has remained unchallenged. The displacement of the Dravidian languages from the north is a historically established fact. This is evident from the residual enclaves of the Dravidian tongues in Baluchistan and southern Bihar (now Jharkhand). For example, Baruhi, Malto, and Kurukh Oraon are living languages of the Dravidian family. The Sinhalese in Sri Lanka is an Indo-Aryan language. It may be argued that the Indo-Aryan languages made incursions into the island country bypassing the Dravidian core. The historical evidence of the southward shift of the Indo-Aryan is not clear. But historians have noted a systematic movement of the Indo-Aryan languages to the south. This process is often referred to as the Aryanisation of the south. We may believe that the mid-Indian region is not exactly a watershed. There is a buffer zone, constituted by the Austro-Asiatic languages, mostly spoken by the tribes.

The story may be reinvented with the help of the available evidence on the rise of Tamil country. There is evidence to show that the Tamil country extended from Vengadam hills to Cape Comorin in the early period of history. It is generally believed that Agastya was among the founders of the Tamil language and grammar. But several historians disagree with this view. In fact, some writings on this subject indicate that the pioneering work *Tolkappiyan* represents the earliest stage in the development of Tamil language. It is also believed that Agastya's work belonged to the period before the sea encroached upon the vast low-lying areas of the Tamil country. The most productive parts of the Tamil country—the coastal plains—were submerged due to the rising level of the sea. The deluge limited the territory of the Tamils and deprived them of their economic base. Some scholars cite the controversy arising from the resistance to the growing intrusion of the 'northern Sanskritic influences'. Panambarnar suggested that the origins of Tamil must have been as old as the eighth or ninth century A.C. There is also some evidence to show that Agastya was related to the Pandayas who were the patrons of the Tamil literature and the Sangam. The Pandayan charters confirmed that the first Tamil power was established and flourished as an Empire. However, Agastya's dominant position in this regard has been challenged. The northern culture, however, penetrated the south and expanded on a slow pace imperceptibly. But the fact remains that there was a southward thrust of the northern Indo-Aryan cultural influences encroached upon the Dravidian fastness in the peninsular plateau region.

The process must have begun at an early date. Whether a composite culture, a mix of the northern and southern traditions emerged or not, is a disputable issue. According to one school of thought the culture and speech of the Tamils was derived from a Vedic seer. The historical references confirm the progress of Aryanisation of the South. However, such penetrations must have taken a very long period of time. Things must have started happening in the beginning of the first millennium B.C. The highly forested Vindhyan mountains and the Narmada trough played a kind of barrier role, obstructing the movement towards the south. The debate does not reject the fact that the southward movement did not necessarily mean overland movement. There was the east coast route, by which the Aryan colonisers seem to have reached the South much before they reached Kalinga. It is noteworthy that the central Vindhyan complex was negotiable at some points, although with difficulty. There were other routes through which the movement took place in phases. It is cited as an evidence that the route from Asmaka to Magadha was just the reverse of the Aryan routes which passed through Avanti (Malwa) and Mandhata. Mandhata is located on the Narmada. It was at this point where the Narmada river was fordable. Paithan was the capital city of Mulaka kingdom. These places, it appears were the earliest colonies of the people from the North. The reconstruction of these routes is obviously difficult, and it is supposed that the passage to the south passed along the Western Ghats. The earliest Sanskritic colonies emerged on the outskirts of Mysore, Coimbatore and Madura. Historians agree that the northerners on their way to the south could have taken a sea route from the mouths of Indus to Gujarat and Kathiawar. There were other routes to reach the eastern coast. Possibly Indo-Aryans reached Sri Lanka by the west coast route. As stated earlier, Sinhalese language is a branch of the Indo-Aryan family of languages.

The process of the Aryanisation of the South was completed during the seventh and eighth century A.C. Moreover, Aryan-speaking people contributed to the emergence of a new synthetic culture. South Indians had maritime links with southwest Asia and these links remained intact for centuries. The linguistic evidence is based on historical records. These links were established because the Indians had trade relations with Syrians and Babylonians. It would not be out of place if one identifies the major items of this trade. Among the commodities exported were gold, spices, and fragrant wood, like sandalwood. It is understood that Indian contact with Babylon was by a sea route. The general assumption is that this must have happened in the fifth century B.C. There are Chinese historical records confirming that the Indians were deeply involved in the maritime trade. They were great seafarers and sailors. The records indicate that these trade links were forged as early as seventh century B.C. It is also indicated that the south Indians were in contact with the Southeast Asian countries and reached as Far East as the Philippines. There is confirmed evidence that the Iron Age implements found in the Philippines were similar to those found in south India.

The history of Vijayanagar Empire is to a very large extent the history of never-ending conflict between the Bahmani and the Vijayanagar kings (Box 5.3).

Box 5.3: Vijayanagar Empire

The subsequent history of that great empire which by resisting the onslaught of Islam, champion the cause of Hindu civilisation and culture in the south for close upon three centuries; thus preserved the ancient tradition of the country in its polity, its learning, and its arts. The history of Vijayanagar is the last glorious chapter in the history of the Hindu south India (Sastri 1981).

But the historian Sastri was confused on the question of relating the Muslims with Islam. When kings clashed with one another they do not define themselves as Hindus or Muslims. Their objectives are very well stated and these are expansion of their territory and a desire to establish the political pre-eminence. There is a clear distinction between Muslims and Islam. The struggle for pre-eminence was not between the Musalman and non- Musalman, it was between one king and the other.

It is noteworthy that the historical records of the Vijayanagar kingdom prepared on the evidence of inscriptions and coins is full of stories of assassinations, court conspiracies, battles, sieges of the forts and the capital cities, agreements between the Bahmani and the Vijayanagar kings. Battles after battles after battles with the Bahmani kings and the Adil Shahi kings continued throughout the history. The territory and the forts changed hands too fast. Centres of Bahmani power like Gulbarga and Bedar emerged and engaged themselves in an internecine war endlessly. Demolition of cities, capture of forts and their relocation were the phases of an unending historical phenomenon. The history is full of intrigues, conspiracies, advances and retreats of armies. The problem was not confined to Andhra or the Vijayanagar kings, it is also recorded that the Vijayanagar armies clashed with the kings of Kalinga. An overwhelming involvement in armed struggles with the neighbours, denied the builders of the Vijayanagar kingdom any time for development activities. But it is noteworthy that despite these adverse factors the ruling dynasty of Vijayanagar kings contributed to the construction of irrigation infrastructure for agricultural development.

Historical records indicate that Harihara was the founder of the Vijayanagar kingdom, which later became an Empire. The administration was streamlined on the Kakatiya model. Following the tradition of the Kakatiya kingdom, Vijayanagar kings organised the country into *sthalas* and *nadus*. Brahmans were employed as *kernams*. A massive programme was launched to bring large areas of land under the plough. Harihara's reign came to an end c. 1357 A.C. and his brother, Bakka, succeeded him. Bakka selected Gutti as the capital city of the empire on the eastern banks located on the northern bank of the Pennar river. Vijayanagar kings were engaged in trade importing goods from abroad. The information based on inscriptions was further supplemented by the travel accounts left behind by the European travellers. The maritime trade in which the Vijayanagar kings were engaged indicates that the trade was an important source of economic prosperity. The trade was the economic backbone of the kingdom. Evidently, a large number of ports were developed to handle the trade. There were ports like Pulicat, Sam, Thome, Poonamallee, Marikanan, Bhavanagiri, Nagore, Pundykottal, Tutticorin, Kanchi, Santhoma and Nellore, to mention only a few.

Other centres of historical importance were Gulbarga, Bedar, Kalyani, Golkonda, Radhakonda, Warangal and Bedham on the Godavari; and Bijapur, Vijayawada, Kunnoor, Velugonda, Kurnool, Devangiri and Narnala on the banks of Tapti and Vodhan south of the Godavari emeged as powerful urban nodes to take care of the administration. While it is true that the Bahmani kings were always at war with the Vijayanagar kings, there were phases in history when the two kingdoms were at peace. A Bahmani king Muhammad II became the king of Gulbarga in 1378 A.C. and the hostilities ended. The Bahmani king was not interested in destructive wars. On the other hand, Bakka's son, Kumar Kampana was far more zealous in this project of expansion of the empire. He, as part of this programme, first subdued the Madura Sultanate. In quick succession the kings of North and South Arcot were reduced. He established his authority in this policy of expansionism. Kampana gathered support to attack the Musalman sultanate of Madura. The war lasted for the period of about five years (1365-1370). As a result of this campaign the Madura kingdom was a Hindu stranglehold. Kampana, however, died in 1374 just three years after this historic victory.

The Vijayanagar kings divided the territory into six *rajayas* or states. *Rajayas* were a sort of administrative units. They were ruled by the princes of the royal family, or in certain cases by the powerful military generals with a good track. These *rajayas* are listed below: (i) Udayagiri (Nellore and Cuddapah districts); (ii) Penugonda (Bellary, Anantpur, and parts of northern Mysore); (iii) Mulvayi (parts of Mysore, Salem, and South Arcot districts); (iv) Araga or Maleha (Banavasi, Chandragutti and Goa); (v) Tulu Rajya also known as Barakur-Mangaluru; and (vi) Rajagambhira.

The emergence of the Vijayanagar Empire was an important achievement of the dynasties of the kings who ruled over the territory. There were at least three dynasties: Sangama dynasty, whose rule lasted from 1336 to 1485; Saluva dynasty ruled during the period 1486 and 1491, and the Tuluva dynasty, whose rule lasted from 1505 to 1576.

While all this was happening in the peninsular region south of the Narmada, Mughals were struggling to establish a vast empire under the Great Mughal, Akbar (1556-1605). The two kings who preceded Akbar—Babar and Humayun—laid down the foundations. A quick recapitulation of the political history of the Vijayanagar Empire may be necessary in order to place the evolution of the empire in its real historical context. Bakka I was followed by his son, Harihara II. His rule extended over twenty-seven years (1377-1404). He consolidated the authority of Vijayanagar with his empire extending over the whole of south India. In c. 1369, Velama was the ruler of Telengana kingdom. A Velama chieftain defeated Kapayanayaka. Bakka II, son of Harihara II was equally interested in expanding the territory of Vijayanagar to satisfy his political ambition. He organised two expeditions around c. 1390. However, the exercise did not produce any results. The expansionist policies of the Vijayanagar kings were also directed against the kings whose territories lay outside the ambit of the Vijayanagar kingdom. For example, the rulers of Vijayanagar annexed to their territory three strategic ports of Goa, Chaul and Dabhol. These ports till now were under the control of Muslim kings. They now became part of the Vijayanagar Empire and the kingdom extended up to Kharapatan and Krishna, which was now the northern frontier of the Vijayanagar empire. Kurnool, Nellore and parts of Guntoor were annexed in just three years time during the period 1382-85. These operations resulted in extending the boundary of the empire to as far south as Ceylon. As it always happens these successes were short-lived. The Turkish kings of the Delhi Sultanate were keenly interested in expanding their kingdom to the region south of Narmada. After the death of Bahmani king Muhammad II, a Turkish slave Tughalchin, ascended the throne. He engaged the Vijayanagar kings in a fierce war against the Vijayanagar kingdom. The war lasted for a long time in the last decade of the fourteenth century. The political history took a new turn. His operations resulted in the reduction of Vijayanagar Empire. However, the armies of the local kings suffered heavy casualties. The Tughlak slave king forced the Vijayanagar kings to sign a treaty for abiding peace. He, however, demanded a huge ransom for the cessation of hostilities. By about the same time, a severe famine occurred in the Deccan resulting in the disintegration of the Vijayanagar Empire.

The expansion and contraction of the Vijayanagar Empire resulted in chaos and disorder, particularly in its peripheral parts. There were two main establishments—the palace and the temple—each having seventy-two departments (*Niyogas*). This establishment also comprised large number of women 'especially chosen for their youth and beauty'. Women were also imported from abroad; many of them were captured in war and enslaved. It is believed that the empire reached its zenith under the *Rajas* of Vijayanagar. The credit goes to Krishnadeva Raya who made a glorious contribution. Vijayanagar kings patronised arts, crafts, language and literature. It appears that through war and peace these kings made their empire the last bulwark of defence against the rising surge of non-Hindu Sultanates.

A Resume

The foregoing exercise in the reconstruction of the historical geographies of the subcontinent has been made to show the evolution of the political and territorial organisation at selected points of time. This approach is described as cross-sectional, meaning thereby that the time has been treated as a constant and the space as variable. The points chosen are important landmarks in the history of the subcontinent.

There are four points of time which have been considered as turning points:

(i) Maurayan Empire in the third century B.C.;
(ii) Harsha Vardhana's Empire, first half of the seventh century A.C.;
(iii) Tughlaq's Empire, early fourteenth century A.C.; and
(iv) Vijayanagar Empire in the fourteenth century.

These empires were selected in order to understand the laws of state formation, their survival and continuity, in the context of the historical processes. Every empire was a form of superstructure and it did not mean abolition of the lower level political structure. The king emperors converted the lower level structure into feudatory states. The king emperor's role was hegemonic. The lower level kings were holding power in their own territory, but they were forced to recognise the empire builders as a suzerain authority. Thus, autonomy and suzerainty remained as on-going processes. This also means that when due to internal weaknesses the empires declined, the feudatory states became all powerful. This was a response to the laws of the land, which were mysterious and added to our confusion.

END NOTES AND REFERENCES

1. There are three different versions of place names in three different languages: Sanskrit, Greek and the Chinese. The Kabul river has been mentioned as Kubha in Sanskrit sources.
2. Chandragupta Maurya's son, Bindusara ruled for twenty-five years. Bindusara's son, and Chandragupta's grandson, Asoka, ruled for thirty-seven years. It may be recollected that Asoka had embraced Buddhism. It was Asoka who sent missionaries of Buddhism to different parts of the eastern world. Asoka's death in 232 B.C. led to the dismemberment of the empire. The Magadhan power was now confined to the Magadha region of Bihar.

6

Society

In this chapter an attempt has been made to identify the broad features of the social structure as evolved in the process of interaction between the communities over time and in space. Space implies that the interactive model is a product of the natural environment, which is the guiding principle of specificity and generality. Our view is that the subcontinent has acquired its characteristic features over centuries of intermixing of ethnic groups. The region displays a baffling diversity in population composition—diversity in physical types, physical appearance of the people, their food ways and dress ways, life-style, belief system and culture. Hundreds of dialects are spoken. There are at least twenty-three languages, which have an established literary tradition. For example, there are *janjatis* and *jatis* with different ethnic origins. There is a caste system, which continues to be a major determinant of social interactive behaviour. It is argued that the social history of the subcontinent by and large is a product of its geographical location.

The Indian subcontinent has inherited a tradition of social hierarchy based on the philosophy of caste. Originally, they were unaware of castes and classes. However, they are conscious of tribal plans and sub-plans. The population also includes tribes and sub-tribes who have a long-standing egalitarian tradition. It has been observed that tracing the origins of the people having significant differences in their ethno-lingual and cultural characteristics is a necessary prerequisite for understanding whether the subcontinent is a pluralistic society or it is an expression of multiplicity of ethnic and cultural formations. It is an oft-repeated statement that different streams of culture groups came from outside and that they occupied different parts of the subcontinent. It is believed that they originated from western and central Asia. These groups came to a land, which offered them multiple opportunities to survive. They were diverse in terms of race. They belong to different stages in their social evolution. For example, India has some 350 tribal communities; they live in clusters. They are confined to their own ecological niches. The Indian subcontinent has offered sufficient scope for the multiple identities to survive. The environment is conducive to their survival and development. Ecologies permit a variety of economic pursuits from pastoralism and nomadism to agriculture and processing of the raw materials from rural cottage industry to factory industry. As already noted, tribes are concentrated in their own enclaves all over the subcontinent. Tribes are not hierarchised but segmented. The environment of the subcontinent has helped the emergence of diversities in their levels of social and cultural development. Many a time this issue has been raised whether the people are 'insiders' or 'outsiders'. One may hold the view that all of them came from outside. Some of them came earlier and some came later. It is an almost established fact that Aryans came towards the very end and their history in the subcontinent is no longer than three millennia, *i.e.*, roughly from twelfth century B.C. to twentieth century A.C. However, there were no insiders because everybody came from outside the subcontinent, be it central Asia, western Asia or regions between the Caspian Sea and the Black Sea. The subcontinent, like other parts of the world, has a long history of peopling. This would mean that the people of different ethnic origins have co-existed in all nooks and corners of the subcontinent. Archaeological evidence shows that these culture streams survived in their own pockets or enclaves across the subcontinent. The basic argument here is that ecologies are variegated and the people have adjusted themselves to these diverse ecological setting. Agrarian communities understandably came to occupy the river valleys with deep alluvial soils and an amicable climate. The agrarian communities adapted themselves to the prevailing environment. The river valleys were surrounded by the hills and the uplands. The denizens of the fringe areas—hills and plateau surfaces—were successful to a very large extent in preserving their original traits. They interacted with the peasant communities but did not get lost in them. They preserved their dialects and the basic elements of their material culture. The first people to appear on the scene were the Proto-Austroloids who were the ancestors of the Dravidian strands. Once upon a time the whole of northern and western India was occupied by the Dravidian groups. Tribal groups, such as the Baruhi-speaking people in Baluchistan (Malto and Kurukh-Oraon speaking tribal groups), indicate that the Dravidian languages once occupied the whole of the subcontinent from Herat and Kabul to the southern tip of the peninsula. The ethnic composition of population changed with the changing times. In the light of this evidence it would be impossible to reconstruct a meaningful story of the linguistic and ethnic history.

Shifting Identities

From People to Communities

The social history of the subcontinent shows that sooner than later the people identified themselves with several categories of ethnicity. These differentiated communities had their own economies. Belonging to a community may be more important than the fluid concept of people. There were communities of speech and communities of faith. The British scholars and administrators in India prepared an exhaustive list of communities. For example, an Indian anthropologist, Roy Burman (1972) told us that there were 350 to 400 tribal communities living in

India alone. The authors of the *Tribal Atlas of India* (1990) concurred with his view. In a recent survey, Singh (1992) listed 2,209 communities living in the country. Earlier to him, Grierson (1903-1928) in his Linguistic Survey of India listed more than 400 language/dialect groups. Each community has an affinity with the certain dialect or spoken language. Tribal groups are recognised as separate entities. There are tribes who belong to a specific area differentiated on the basis of their ethnicity and dialect. The very name of a tribe suggests a possible relationship with a specific language. The British anthropologists and administrators attempted to understand the ground reality of their geographical distribution. The hills were named after the tribes and the tribes after the hills. For example, Garo, Khasi and Jaintia hills were named after the specific tribal groups living in this area.

From Communities to Nationalities

It may be argued that the ethnographic history of the subcontinent reflects the fact that we have multiple identities. With the passage of time the speech communities became more and more exclusive and soon identified themselves as religious communities. These formations were reinvented when the idea of the two-nation theory was advanced in the pre-partition period in the name of religion. While categories, such as Hindu, Muslim, Sikh and Christian, are based on religion, they are not mutually exclusive. For example, a Hindu may belong to a culture group like the Kashmiris. When they embraced Islam they did not delink themselves from the Kashmiri tradition. It does not matter whether you are a Hindu or a Muslim or a Sikh or a Christian because you are attached to a region, which becomes the basis of your ethnic identity. When people change their religion, they retain their pre-conversion caste names. They are Kashmiris first, and Muslim or Hindu later. The idea of Kashmiriat is transcendental to religious divisions. The same statement applies to all other ethnic groups of the subcontinent. They may be Bengalis or Assamese, Oriya or Marathi, it does not matter. The identity of the ethnic groups depends on their social history and exposure to a social environment. The politicisation of the idea of religious identity led to the emergence of a nation based on the now discredited universal principle of one people, one language and one nation. However, the people of the subcontinent never lent themselves to such a division. The architects of the two-nation theory did not realise that in the history of the subcontinent, territory was never divided in the name of religion. The Eastern Wing of Pakistan separated from the Western Wing in the name of Muslim Bengali culture. The birth of Bangladesh was the most powerful indicator of rejection of the idea of communities on the basis of religion. In fact, religion alone cannot be a basis of nation-building. It cannot constitute or reconstitute the territory. In a way, the single nation was once again partitioned to preserve the Bengali Muslim identity. In the wake of struggle for independence of the East Wing of Pakistan, the Indian leadership indirectly accepted the three-nation theory. The Baluchis of Pakistan are also reported to be struggling for a separate nationhood. The model of East Pakistan is very much in their mind.

Society and Social Formations

Society or *Samaj*[1] is a phrase not exactly applicable to the people of the subcontinent. All other things remaining the same, the people of the subcontinent are not conscious of their being a member of an integrated societal formation, which is larger than their own particularity. There are other definitions of their identity, such as caste, class, community, culture, tradition, and the contemporary reality. These variables define their status/rank in the social order.[2] But the fact is that all the criteria mentioned above are divisive. We hear of Arya Samaj or Brahmo Samaj as equivalent social structures in other religion-based denominations. Structures, such as Khatri Samaj are more understandable than any other form of aggregation on the basis of social and cultural categories. The subcontinent is a peculiar case: people are recognised on the basis of their loyalty/assertion to an exclusive social formation. They live in their own shells. In fact, there are shells within shells; the social groups meet and circulate among themselves; celebrate festivals among themselves, and this is done along caste, creed and community lines. There is no concept of a holistic society, which cuts across the established categories of caste, community and culture. The problem is further complicated when one focuses on the state of village society. It is a well-known fact that the rural masses are divided in the name of caste and class. The consciousness of rank, occupation and class within the caste society is more important than any other criteria. For example, there are caste councils, particularly in rural areas, which occasionally meet to discuss the issues that crop up at a certain time in history. In these matters a definite position has to be taken by the people collectively. One such issue is the support to be extended to a particular political party on the eve of general elections. The council, it is understood, decides their options or action plans. For example, the Jats and the Gujars of Haryana, and western Uttar Pradesh assemble at a centrally located village in the region. These are large conclaves, *bhandaras* as they are locally known. These meetings are attended by the members of the same caste groups distributed in geographical space over a vast area often consisting of hundreds of villages.[3] The political parties very often field their candidates in consultation with the leaders of the other caste groups in the same region. Candidates for the general elections to the parliament, or the state legislative assemblies, are selected in consultation with the elders belonging to different caste groups of more or less equal ranks. Evidently, political decisions are taken collectively and not individually. Indian polity is described as a democracy, even though individuals are not always free to exercise their right of franchise independently. On such occasions several castes meet in a centrally located village and take decisions collectively in consultation with the other castes living in their neighbourhood. Several decades ago a combination of dominant castes launched a joint political party in north India. It was nick-named as AJGAR (Ahirs, Jats, Gujars, and Rajputs). It is interesting to note that AJGAR emerged as a formidable political force to reckon with. It appears that in Indian elections the rallying factor is neither the political party nor the individual. The rallying factor is the caste identity.[4]

Social groups, and not individuals, often take rigid positions in matters of marriage, social interaction, rank and attitudes to commensality.[5] An eminent Indian writer, Nirad C. Chaudhari, once described himself as a Bengali Brahman and not an Indian. Here, being a Brahmin is more important than being an Indian. Paradoxically, Indian polity is democratic, and yet the people and the parties by and large have little idea of democracy. One wonders whether the mushroom growth of political parties as in the subcontinent really means democracy. In fact, it makes a mockery of democracy. Each individual leader is an institution in himself/herself, launches his/her own party. The media and the party men often describe him/her as a 'supremo'. One does not come across such definitions in the political party system of either Britain or the USA. There is some substance in the fact that both the United Kingdom and the United States of America have a two-party system. This ensures the proper functioning of a democracy.

The concept of *Samaj* is confusing. In the context of the subcontinent, society means that it is a parochial, ethnically defined and communal formation. The society is viewed as an ambiguous structure. It is not a secular[6] concept by any means. Evidently, society implies the people as a collectivity. Both India and Pakistan view the social groups within their population as hierarchised and structured. They belong to different strata based on caste, class and *jirga*. This is because the social norms—traditions followed, spoken dialects/languages, religious practices—differ substantially among the different strata of the society. As already observed, dress-ways differ. Even the culinary preferences reveal a vast diversity in different parts of the subcontinent.

Ethnic Antecedents

Settlement on land was not a necessity for the early human groups. Settlement was preceded by a long phase of nomadism/nomadic pastoralism. The process of sedantrisation began with the emergence of agriculture as a mode of production. For millennia of years human population remained dependent on hunting and food gathering. It is believed that the transition from nomadic state to partially settling down on land took a very long time in history. Earliest records of settlement of the peasant communities are dated as seventh or eighth millennium B.C. Archaeologists have dug out the oldest village sites in Iraq, Palestine and Turkey. The evidence shows that the early people adopted agricultural mode of production before settling down permanently. These primeval communities practiced a form of primitive agriculture, sometimes also referred to as shifting agriculture. Their agricultural mode may be described as crop husbandry. But these groups did not give up nomadism altogether. They were partly settled and partly nomadic. Perhaps they were practising a form of shifting agriculture (known as *jhum* in the eastern parts of the subcontinent). They selected food grain crops found in the wild state, such as wild wheat and wild barley. In so far as the subcontinent is concerned, this phase is generally dated as the middle of the fifth millennium B.C. Baluchistan village sites are the oldest in the subcontinent. Some archaeologists believe that early beginnings of agriculture notwithstanding, the process of diffusion took several millennia before the agricultural mode reached the Indo-Iranian borderlands in Baluchistan. Several millennia also passed before the agricultural mode reached the Ganga valley in north India. Likewise, there is evidence to show that paddy cultivation in eastern India and other parts in Southeast Asia, such as Malaysia and Indonesia, were by and large older than the Baluchistan sites. However, an agriculture-based civilisation comparable to the Indus Valley did not develop in eastern and northeastern India where paddy was the main food grain crop. On the other hand, wheat and barley culture culminated in the Indus Valley and eventually led to a civilisation by the middle of the third millennium B.C.

Archaeological records show that the sedentary peasants were riverbank communities. They were settled only at places where it was possible to divert the river waters to the fields on the riverbanks. In so far as the question of time is concerned it is noteworthy that the gap between the Indus Valley and the Ganga Valley civilisations was no less than a thousand years in time and a thousand miles in space. While the archaeologists stand on their feet firmly because of the evidence unearthed by them, ethnologists are not so sure-footed. This means that associating the people with a certain culture is bound to be a difficult exercise. Perhaps one can make intelligent guesses. But the guesswork is not an easy exercise either and cannot be described without ifs and buts.

Who were the first people to reach India? And when did they arrive? Their origin, as far as we know, is shrouded in mystery. This question can be answered by examining the extant evidence, both racial and ethno-lingual. Admittedly, the question of race has always been vague in the history of the subcontinent. Several writers have suggested that the earliest people who came to India were the Proto-Austroloids.[7] However, it is generally believed that earlier to them a stream of migration had brought the Negrittos to India. The general belief is that the Proto-Austroloids were far more numerous and gradually the Negrittos were assimilated into the Proto-Austroloid stock to such an extent that they almost lost their identity and became extinct. It may be posited that the present population of the subcontinent has been derived from different ethnic groups who came to this part of the world at different points of time. They, along with the Palae-Mediterraneans and the Mediterraneans were the builders of the Indus Valley Civilisation some 3,000 years before Christ. The story of the agrarian civilisation started with Neolithic cultures in the Indo-Iranian borderlands in Baluchistan.[8] The pre-Harappan sites in the Indus Valley, more particularly in Sind,[9] now in Pakistan, provide enough evidence to confirm this conclusion. The Neolithic cultural tradition drifted eastwards and was first discovered in Pre-Harappan sites in Baluchistan and Sind. It gradually spread to occupy the entire Indus Valley. It later shifted to other parts of the subcontinent, particularly the Saraswati and the Ganga Valleys. According to this view, the peasant communities gradually covered the whole Indus Valley in Sind and neighbouring Punjab. Perhaps this region became the main theatre of the ethnic inter-mixing leading to the emergence of an agrarian civilisation, the first ever in the world after Mesopotamia and by and large contemporaneous to it. All evidence points to the fact that the civilisation lasted for at least fifteen hundred years up to the middle of the second millennium B.C. The skeletal remains found during the excavations in the cemeteries of Harappan cities indicate that the Mediterranean stock were partners in this venture of building a riverine civilisation. These groups reached the subcontinent through the difficult passes in the Hindu-Koh Himalayan mountain complex as well as after crossing the passes in the Sulaiman and the Kirthar ranges, bordering Sind. On the other hand, while all this was happening in the northwestern region of the subcontinent, various strands of the Mongoloid stock—broad-headed (*Brachycephalic*) groups as well as long-headed (*Dolychocephalic*) groups—appeared in the northeastern region of the subcontinent. There are passes all along the Himalayan range from Indus to Brahmaputra.[10] There is evidence to show that the people having affinity to the Mongoloid stock remained by and large confined to the Himalayan and the sub-Himalayan regions from Laddakh to Arunachal Pradesh and Tripura. The lowly hills in the northeastern region of the subcontinent bordering the Indian states of Nagaland, Manipur and Mizoram were negotiable. The tribes crossed these passes and entered the subcontinent from the neighbouring parts of Myanmar. However, they remained confined to the northeastern region, including the Assam Valley and the surrounding neighbouring states. It has been suggested that their presence in the subcontinent resulted in intermixing with other racial stocks. In this connection reference may be made to the Ahoms. The geographical distribution of these Mongoloid groups shows that they were there in the region on either side of the Indo-Burmese border. Like the Indus, the Ganga and the Brahmaputra Valleys became a theatre of racial intermixing, which lasted over long periods of time. After the entry into the subcontinent, the Mongoloid groups did not exactly settle down but remained on the move. The ethno-lingual diversity of the present-day population of the northeastern region of the subcontinent, as far as we understand it, is a result of ethnic intermixing.[11] Moreover, a hybrid variety of ethnic groups emerged in other parts of the subcontinent as a result of the Mongol invasions in northwestern India. Thus we can say that these groups spread over a large area outside the sub-Himalayan and northeastern parts of the subcontinent. The Mughal women may be described as the best examples of a synthetic race. In fact, the Indian women became a model of beauty,[12] perhaps the most beautiful outside Azerbaijan and Iran.

Writings on South Asia generally emphasise the partially enclosed character of the subcontinent. In fact, it has been over-stated. The subcontinent was never closed to the wandering people. The truth is that the subcontinent attracted[13] not only the pastoral nomads but also the armies of the empire-builders like Darius of Iran and Alexander of Macedonia. The Achaemenid and the Greek Empires extended over a vast geographical area. They virtually incorporated the entire northwestern region from Kashmir to Gujarat and from the Oxus to Yamuna. The Achaemenid rule lasted for two hundred years, which is not a small span of time. A Greek prince, Alexander, overtook the territories of Darius. Evidently, Iranians and the Greeks left behind their imprints on the people, more particularly the ruling elites, in the subcontinent. For example, the idea of commemorative pillars such as those of Ashoka was derived from Darius, who had raised similar pillars in various parts of his vast empire. There are layers over layers of different racial groups who very often merged with one another and created new hybrid varieties.[14]

Several anthropologists, ethnographers and historians hold the view that the evidence of skeletal remains of the Early Man in South Asia is missing. This fact, however, disturbs a section of the Indian nationalists[15] and specialists. They do not subscribe to the view that all of the present racial stocks came from outside and that no one is an insider. A section of Indian historians, archaeologists and ethnologists do not think that the existing evidence suggests any such conclusion. The conclusion is based on the assumption that the earlier biological evolution of the hominids did not take place within the subcontinent. In fact, intermixing of different ethnic groups contributed to a process of racial synthesis. It is difficult to say particularly, in the context of large scale intermixing, that any pure races are still there in any part of the subcontinent. A broad uniformity in the physical form of our people does not accept the idea of a pure race. The idea of pure race is not acceptable, even though some groups in isolated parts of the subcontinent, such as Kashmir, Ladakh, Baluchistan, Northwestern Frontier Province of Pakistan, northern Nepal and Afghanistan have retained their original racial traits by isolating themselves from the rest of the population. Perhaps the river valley settlers squeezed the primeval groups out. These people came from different parts of Asia and Africa particularly, from the regions, which became nodes of human centres of dispersal in history: Mongolia, North China, Trans-Oxania (earlier known as Turan), and the southwestern margins of Asia, Ethiopian plateau in Africa, southeastern Europe (Georgia, Romania, Armenia and the Dinaric Alps). They found themselves in a land, which was rich in natural resources. They created niches for themselves. Intermontane valleys in the Hindu Koh-Himalayan region accommodated these groups according to their own preferences. Thus the diversity in ethnic structure and ethno-lingual background is the product of ecological diversity observed in the sub-Himalayan region from Indus to Brahmaputra.[16]

Anthropologists and archaeologists have made a substantial contribution to the quest for Early Man in the Indian subcontinent. They have found the early Palaeolithic tools belonging to the Pleistocene epoch. This period was marked by the Ice Ages, which advanced and retreated in succession. Paradoxically, one fails to find any association between human skeletal remains and the artefacts left behind by the early human groups. Evidence is available in the form of stone tools of different sizes in different parts of the subcontinent. Along with the tools, pottery remains (potsherds) have also been dug out from the different sites in the subcontinent. It is known that the Neolithic cultures marked a kind of transition from nomadic pastoralism to primitive agriculture (Appendices III and IV). Diversity in languages and dialects found in various parts of the subcontinent throws some light on the ethnic antecedents of the people. But the evidence is not sufficient. However, some conclusions have been drawn from the materials so far discovered. B.S. Guha (1944) made an attempt to identify the patterns in the geographical distribution of racial groups[17] as classified by him. However, Guha's scheme of classification of ethnic/racial groups of the subcontinent can best be described as a generalisation, perhaps a simplification. Differences of opinion exist between anthropologists and archaeologists on the question of classification of ethnic groups. More difficulties arise when an attempt is made to trace the routes taken by different racial groups who came to India and were found over a vast geographical area. These studies have not challenged the substance of Guha's scheme of classification, although differences of views exist even today. It is also believed that European studies on racial issues may be the basis for a renewed look at the whole question.

Race and Caste

There is a general feeling of discontent among the South Asian scholars on the issue of the origin of the immigrants and the source regions from where they came. A certain school of thought contended that all of them came from outside the subcontinent. This view has also been expressed that India was the home of the Indo-Aryans and that they migrated to other parts of the world, such as Iran, West Asia and Eastern parts of Europe. But this is a controversial issue. Archaeological findings reveal the inherent weaknesses and the inconsistencies of this logic. *The Historical Atlas of Iran* (Tehran University 1971) schematically describes the routes by which the early Aryan groups came. This evidence is conclusive to prove that the Indo-Aryans came from outside the subcontinent. The Achaemenidian Empire included the whole of Afghanistan, Karman, Pars or Parasia, Arachosia, Sogdiana, and Bactria.

Some Indologists have tried to link caste with race. They observe that the basis of the caste was the Varna system. Hierarchy was based on colour of the skin and not the occupation. However, the whole question of caste needs re-examination. The general view is that with the passage of time, caste became more and more 'occupational', meaning thereby that the occupations were related to the caste groups. Each occupation was exclusive in itself. The word caste implies the division of labour and affiliation to some occupation. Thus, distinction is made on the basis of work, clean or unclean. Caste became the basis of social stratification. Those who care for the dead bodies of men and livestock are supposed to be placed at the bottom of the caste hierarchy. Similarly, those who defend the territory from inside or outside enemies are given a high rank in social hierarchy. The obvious example is of the Kshatriyas. Naturally, when occupations multiplied, a kind of specialisation in a certain department of work became the basis of further caste formation.

Nesfield (1855) argued that the race was not the basis of caste. However, Risley (1891) pleaded for a scientific basis for the classification of caste. Using anthropometry as a method, he asserted that one way of addressing this issue might be the measurements of the nose, face and head. These indices were developed on anthropometric criteria, such as facial index, nasal index and the cephalic index. The facial index was based on data in numerical terms showing the length of the face from the forehead to chin and breadth from temple to temple.

Similarly, the nasal index was calculated on the basis of the length of the nose and its width; the cephalic index was computed on the basis of measurements from the forehead to the occiput and width from temple to temple. Risley's voluminous work on tribes and castes of Bengal (1891) and later his work on the people of the Indian subcontinent in other provinces of British India (1915) could be a viable basis of classification. Risley suggested that specimen population of the high and the low caste should be measured and then placed in a certain stratum. Anthropometric data for two categories of caste, a high-caste group such as the Brahmins and a low-caste group such as the Chamars, was collected. Interestingly, the two caste groups are not uniform in physical type.[18] They are worlds apart in terms of hierarchy and lineage. Ghurye (1950) held the view that the anthropometric data would be of no use if we take into account the racial mixture as well as the racial uniqueness. His method of study was basically dependent on the general knowledge about the ethnology of India. In the same series of studies, data were collected through fieldwork for the Kurmi and the Musahar castes of Bihar, Bhils of Khandesh, the Kayasthas and the Brahmins of United Provinces of Agra and Oudh (now Uttar Pradesh). For example, Brahmins of Uttar Pradesh were supposed to be the representative type of the ancient Aryans.[19] Data were also collected for the Chuhras, a lowly caste, associated with leatherwork and the Khatris (Chhetris) of the Punjab. Comparisons between the Uttar Pradesh Brahmins and the Chuhras led to the conclusion that the Brahmin of Uttar Pradesh is as much akin as the latter is to the Khatri of the Punjab.[20] Another conclusion was that the Uttar Pradesh Brahmin has essentially retained the same physical traits as that of Punjabis and ancient Aryans.[21] We now find that the Kurmis are close to the Brahmins while the Chamars and the Doms stand much differentiated from the Brahmins. Studies conducted in Bombay Presidency indicated that the Deshast Brahmin bears as close an affinity as to the Son Koli, a fisherman caste, as to his own compeer, the Chitpavan Brahmins. The Mahar, an untouchable caste of the Maratha region comes next together with the Kunbi, the peasant caste. In the same order follow the Shenvi Brahmin, the Nagar Brahmin and the high caste Maratha.[22]

Scholars working in this area have suggested an order of ranking on the basis of data on the castes of United Provinces of Agra and Oudh. Table 6.1 presents a comparative picture of caste ranking. Evidently, we cannot say that the anthropometric data on caste are final and conclusive.

Guha was convinced that the Negrittos and the Proto-Austroloids lived together and came to the subcontinent before or during the Early Palaeolithic times. While these groups came by the northwestern routes, northern and eastern parts of the subcontinent received streams of the wandering people of the Mongoloid stock. It is not clear whether the Mediterraneans came at the same time as the Proto-Austroloids. Even though their presence in India is established by the skeletal remains dug out from the cities of Harrappan civilisation (Guha 1944: 18-19). It is understood that the Mediterraneans came in different streams. They had striking variations in their racial elements. This was due to their origin in different parts of Western Asia and the adjoining Mediterranean lands. Gradually, and in different spans of time, the subcontinent received broad-headed (Brachycephalic) people like Alpinoids, Dinarics and Armenoids. The Nordic stock, represented by the Indo-Aryans, appeared in some parts of the northwestern region of the subcontinent.

Table 6.1: Ranking of Castes based on Anthropometric Criteria

Rank Order	*According to Risley and Others*	*Rank Order*	*According to 1901 Census of India*
1.	Babhan	1.	Brahmin
2.	Brahmin	2.	Babhan
3.	Chhatri	3.	Rajput
4.	Kayastha	4.	Kayastha
5.	Kurmi	5.	Bania
6.	Tharu	6.	Kurmi
7.	Bania	7.	Lohar
8.	Bhar	8.	Mallah and Bind, etc.
9.	Kol	9.	Bhar
10.	Lohar	10.	Tharu, etc.
11.	Dom	11.	Kol
12.	Pasi	12.	Muchi, etc.
13.	Musahar	13.	Pasi and Doodh, etc.
14.	Chamar	14.	Chamar, Musahar, and Dom, etc.

Source: Table prepared by the author on the basis of studies conducted by the Anthropological Survey of India.

It is believed that the western Brachycephalic groups came sporadically and in several streams. There were three main routes by which they came to the subcontinent. The first stream passed through Baluchistan, Sind, Kathiawar, plains of the Mahi and Sabarmati rivers in Gujarat, and ended in Tamil Nadu. The second stream passed through the northwestern region of the subcontinent, taking a route by and large parallel to the Siwaliks. Its eastward movement ended in the deltaic region of Ganga. The third stream passed through Chitral, Gilgit, other parts of Kashmir and ended in Nepal. Historians hold the view that they reached India sometime in the middle of the second millennium B.C. The northwestern parts of the subcontinent have yielded sufficient evidence to show that they came from Iran, or Aryan. This conclusion is based on the available linguistic and ethnic records. This racial group is found in the North West Frontier Province of Pakistan and undivided Punjab. On arrival in the subcontinent, they found themselves in the midst of other racial groups who came before them. They were distinguished on the basis of their skin colour. It appears that they became conscious of their racial superiority. It was this consciousness that became the basis of the *Varna* system. Sufficient evidence is there to show that the colour of skin was a critical factor in classification. In the primordial stage when caste consciousness became a common feature, they assigned themselves the highest rank in caste status. The other ethnic groups of diverse origin, such as Scytheans and Indo-Scytheans, Sakas and the white Huns came intermittently. In the same sequence northwestern region of the subcontinent was exposed to an organised Mongoloid invasion. Today the subcontinent presents a remarkable mix of the people (races). Centuries of living together have led to a distinctive physical type who is typically Indian.

The population of the South Asian subcontinent today is an end-product of millennia of living together and ethnic intermixing in the subcontinent. While there are diversities in ethno-lingual characteristics and communal formation, they are like the shades of the prism. The underlying unity is as important as the overlying diversity. The progress of Indian civilisation conforms to the basic elements of the diverse ecologies and social organisation—concentration of the peasants in the river valleys and the pastoral nomads in the periphery of the basins. It may be recapitulated that the river valleys were more attractive to the agrarian groups. The emergence of nuclear regions (a la Subbarao) was obviously a post-agricultural phenomenon. However, before the advent of agriculture, hunters and food gatherers were not interested in the river valleys for any particular reason. They wandered from place to place in quest for food. India presented diversity in its ecological setting to such an extent that the tribal, agrarian and pastoral modes could survive in a neat spatial arrangement. While the agrarian mode spread horizontally in the river valleys, the inter-fluvial tracts were favoured by the nomadic pastoralists. Higher up on the plateau surfaces or in isolated hills and mountain ranges tribes lived in their own exclusive world.[23]

Regions and Regionalism: Subbarao's Schema

A noted Indian archaeologist, Subbarao, developed a highly schematic view of regions and regionalism. But his approach was schematic.[24] The idea is to describe the structure of the society. While the political boundaries, which divide the people of South Asia, are regarded as sacrosanct, they do not and cannot be considered as permanent. The subcontinent presents a case of shifting boundaries. They have been shifting in the past and would be shifted in the future. This is what the history of the subcontinent shows. The external boundaries are guarded jealously, so much so that any violation of these imaginary lines is considered a culpable crime for which one can be booked for a term in the Indian or Pakistani prison. The reports come almost daily that the fishermen drifted into each other's territory and are punished for that imagined violation of law. Boundaries are not visible on the sea.

River valleys were described by Subbarao as the perennial nuclear regions. While it seems to be a good characterisation, it is difficult to accept his idea of regions of attraction, regions of repulsion, *cul-de-sacs* and round-abouts. There were several round-abouts and several *cul-de-sacs* in the subcontinent. The Z-pattern of the routes of migrations within the subcontinent is one of the alternatives. At best it is a simplification of what happened in history, in fact, a generalisation. There are other routes to be identified. The people who came to the northwestern parts of the Punjab were not essentially taking the river valley routes. Movement in the Indo-Gangetic divide region was not obstructed by the lowly Aravalli Hills. One can work on another set of hypotheses. In fact, there were several other choices open to the early wanderers. Streams of migrating people took several routes. For example, the northern route brought the people to a point where the Yamuna was fordable. The lower parts of the Ganga-Yamuna *doab* were accessible by the Ganga valley route to reach the Bihar plain. If one chooses the northern route past Yamunanagar one reaches the Upper Ganga-Yamuna *doab*. The diversion from the north Indian plain to the south via Madhya Pradesh and Gujarat was not the only option. Take for example, National Highway No. 8, which connects Jaipur with Mumbai, is a testimony of the fact that relief features do not pose a serious problem in intra-subcontinental movement of the early people. The Godavari valley route to the eastern parts of the Godavari-Krishna deltaic tract and the eastern littoral region also appears to be one of the alternatives. These are big issues. They can be addressed by the archaeologists and the historians, if not by ethnographers and the language historians. In this context the significance of the Makran coast route cannot be underrated. One can take this route to reach Gujarat and from Gujarat to reach Baluchistan or from Indus delta in Pakistan to Kathiawar and the northern parts of northern Gujarat. History of penetration of ethnic groups and their redistribution in the geographical space of the subcontinent can be reconstructed by using archaeological and literary evidence together. In situations where archaeological record is missing, literary sources have to be used but with care. A review of the social scene is necessary in order to recognise the basic waft and web of the social fabric. In the following paragraphs an attempt has been made to portray the social structure as it is observed today in the perspective of time.

Regions Based on Ethnicity, Culture and Dialect

Regions and Regionalism

The diversity in physiography and climate, as described in Chapter Two, has resulted in the corresponding diversity in cultural traits, with each culture-group finding a place for itself. They live in their own enclaves surrounded by wilderness. However, on-going development projects have reduced the area of wilderness. Ethnicity and dialect/language are the main criteria on the basis of which ethnic regions of the subcontinent can be identified. These factors have operated through the ages and define by and large the identity of the people of the subcontinent. A study of the ethnographic data for the states/provinces of the Indian Union as well as Pakistan, Bangladesh, Nepal and Sri Lanka leads us to the conclusion that the people belong to distinct culture groups and that their dialect/language is the best expression of their cultural identity. This identity is an expression of the regional consciousness, which often expresses itself in regionalism. The idea of regionalism carries with it a sense of separatism, implying that regions are different from the nation as a whole. But it may be disputed that the nation is larger than the sum of regions. The controversy of region *versus* nation is very old and India after independence opted for federalism as the basic principle of polity.

Table 6.2: Regions based on Ethno-lingual Identity of the Subcontinent

Macro Level	*Meso Level*	*Micro Level (Spatial Distribution of Speech Communities)*
Dravidian	South Dravidian	Tamil*, Eurokala, Malyalam*, Yerava, Kannada*, Kuruba, Coorgi/Kodaga, Tulu*, Koda/Kota.
	Central Dravidian	Telugu*, Vadaga, Kui*, Kolami, Gondi*, Parji, Koya*, Khond*/Kondh, Konda.
	North Dravidian	Kurukh-Oraon*, Malto*, Brahui.
Indo-European	Dardic	Dardi, Shina, Kohistani, Kashmiri*.
	Indo-Aryan: Northwestern	Lahnda, Seraiki, Sindhi*/Kachchhi.
	Indo-Aryan: Southern	Marathi*, Konkani*.
	Indo-Aryan: Eastern	Oriya*, Bihari (Maithili*, Magadhi*, Bhojpuri*), Bengali*, Assamese*
	Indo-Aryan: East central	Awadhi*, Bagheli*, Chhattisgarhi*
	Indo-Aryan: Central	Hindi: Western Hindi: Bangaru, Braj Bhasha, Bundeli*, Kannauji; Urdu*, Punjabi*/East Punjabi; Rajasthani:* Marwari*, Mewari*, Mewati*, Kurmali Thar*, Bagri*, Banjari*.
	Indo-Aryan: Northern	Eastern Pahari/Nepali*, Central Pahari: Kumauni*, Garhwali*; Western Pahari: Jaunsari, Sirmauri*, Kului*, Mandeali*, Chameali, Bharmauri/Gaddi, Churahi.
Austric	Monkhmer	Khasi*, Nicobarese.
	Munda	Santali*, Munda, Mundari*, Bhumij, Birhor, Koda/Kora, Ho*, Asuri, Brijia, Korku*, Khaira*, Juang, Savara*, Gadaba, Kharwari.
Sino-Tibetan	Tibeto-Himalayan: Bhotia Group	Tibetan, Balti, Ladakhi, Lahauli, Sherpa, Sikkim Bhotia.
	Sino-Tibetan: Himalayan Group	Kinnauri, Lepcha, Lahauli.
	North Assam: Nefa Group	Aka, Dafla, Abor/Adi, Adi*, Miri*, Mishmi, Mishing.
	Assam-Burmese: Bodo Group	Bodo*, Kachari, Lalung, Dimasa, Garo*, Koch, Rabha*, Mikir*/Karbi, Tripuri*.
	Assam-Burmese: Naga Group	Angami, Sema*, Rengma, Khezha, Ao*, Lotha, Kabui, Mao, Tangkhul*, Konyak*, Pochury, Phom, Yin Chungre, Khiemnungan.
	Assam-Burmese: Kuki-chin Group	Manipuri*, Meitei, Thado*, Paite, Lushai*, Rangkhul, Halam, Hmar.

* The asterisk sign shows that the numerical strength of population speaking a certain language group is more than 100,000.

It may be noted that the spatial distribution of dialects and languages has played the basic role in the formation of the regions. Regions give rise to regionalism, which in the context of the subcontinent results in a kind of association with a certain unit of territory and it is likely to lead to unity within the region. It is a complex phenomenon and it cannot be expressed in hard and fast rules. This is also a fact that the spatial concentration of a certain ethno-lingual group generates regional consciousness, which in situations can be divisive, if not secessionist.

Geographical location of culture-groups cannot be ignored in nation-building. Unlike France, Germany or Spain, the subcontinent is not a uni-lingual structure. It may lead to the assertion of the group to be recognised as the basis of its identity. For example, the Seraiki language-group in Pakistan is currently 'demanding' the recognition of their identity and their demand is that Pakistan's Punjab should be divided to constitute a new province of Seraikistan. If we examine the political situation in Baluchistan the whole evidence points up to the possibility of the province emerging as an independent autonomous region/country on the pattern of Bangladesh. Regionalism may also be a ploy to protect the interest of the group to combat with the federal government in case if it has taken certain steps detrimental to the fulfilment of the regional aspirations. In India after independence, language-based regional consciousness became the criterion for state formation. The people are conscious of the fact that their cultural roots lie in the language they speak, and the literature they produce in that language. For example, the Partition of British Empire in the subcontinent in the name of religion proved to be a fragile basis. It was realised that language and ethnicity, not religion, were more viable indicators of identity. East Pakistan separated from West Pakistan on the same basis. It may be recapitulated that the battle between the two wings of Pakistan was really the battle between the Bengali and the Punjabi speech communities and the cultural realms they belong to. The demographic concentration of the Musalmans in certain parts of the subcontinent was the main factor that led to the emergence of Pakistan. But the same principle could not keep the two wings of Pakistan together.

Table 6.2 shows the scheme of regions based on ethno-lingual identity.

Culture Realms

Pakhtoons/Pushtuns Culture Realm

In this section we have tried to present a geographical distribution of different ethnic groups living in the subcontinent. People with a distinctive cultural background are clustered/concentrated in their enclaves, even though it may not be possible to delink them from the general socio-cultural scenario. It will be in the fitness of things if we start the story from Afghanistan, a perennial roundabout in the history of the subcontinent.

With a total population of 22 million, Afghanistan presents an interesting mix of ethnic and racial groups. About one-half of the population (52 per cent) consists of Pathans believed to be the true Afghans. The Afghans are further divisible into two major clans—Durrani and Ghilzai. The social order is based on the tradition of *Jirga* formation, which is still intact. While *Jirga* is a classical model of ethnic and cultural identity based on the concept of blood relationship or an extended kinship, it is a typical example of the clan identity and territoriality. Other ethnic groups found in the country include the Tajiks (Iranian stock), Hazaras and Uzbeks. The Tajiks constitute about 20 per cent of the total population, Hazaras 19 per cent and Uzbeks 9 per cent. There are about 2.5 million nomads who remain on the move, depending on the weather conditions. An estimated 5 to 6 million Afghans live as refugees in the neighbouring countries of Iran and Pakistan.[25] They also live in several cities of north India. Keeping in view the political instability caused by successive interventions from outside, the refugee problem can never be resolved. Throughout the recent history, the situation has remained fluid; first, the British-Afghan wars before 1947, then the Russian occupation of Afghanistan for fifteen years (1975-1990) and finally, the allied US-UK invasion of the country in 2001. These events have kept the people on the move. The people of Afghanistan have suffered one setback after the other.

Located astride the pathway to the richer and more prosperous parts of the subcontinent in the east and the south,[26] Afghanistan's history is full of references to the episodes of intrusion and subjugation by the expanding empires: first, by the Achaemenid[27] Empire and the second by Alexander the Great. The human/ethnic constituents changed with each incursion or invasion. The Kabul valley has acted like a funnel, pouring people with different ethnic origins into the subcontinent. The remarkable point is that this continuing process of incursions started since the very early stage of Palaeolithic times. The role of the Kabul River in this process of incursion/invasion and the settlement of the primeval communities cannot be overstated. Kabul valley has been a meeting place, and a roundabout, but not a melting pot. Iranian imprints on the Afghan soil are too obvious to miss.[28] The corridor role is also evident from the rich and diverse cultural heritage of the Afghan people. These influences came as layers over layers of the Iranian civilisation followed by the Hellenistic tradition, Buddhism[29] and Islam.[30] These ingredients could not contribute to a synthesis. In the context of Afghanistan one cannot say with confidence that ever in the history of peopling a stage came when a composite culture emerged. The population consists of clans, not people. The *jirga* is the basis of identity. The territory or the administrative units do not matter.

Afghanistan today is in shambles. It has a long history of civil strife and outside intervention. The symbols of fear are largely writ on the faces of the Afghan people. It is a country of devastated homes and broken hearts. The US-UK allied forces in their zeal to eliminate vestiges of the Taliban kind of governance did not spare even the mosques, hospitals, schools and the wedding parties. For the carpet bombers, places and people are not important. The operations of the allied forces were aiming at voluminous and total destruction. This was a show of brute military power of the most developed nation on the planet Earth.[31] The war was imposed on them, they did not opt for it. Relieving the Afghan mind from the agony of war would not be easy, even though the Afghan children are accustomed to the jarring sound of air raid sirens and the resounding sound of gunfire. The people did not anticipate that they were facing nothing short of genocide.

Much of the present-day Afghanistan is what the circumstances have made it. The people of Afghanistan waged an armed battle to throw the Soviet occupation army out. However, the US extended all possible support to the Afghan guerrillas who were involved in an anti-Soviet retaliatory action. Unfortunately, the ouster of the pro-Soviet regime did not bring peace. The US support to the Afghan guerrillas in those eventful years resulted in the expansion of the insurgent movements. One can look back in time to find that the leader of the Afghan guerrillas became a hero. It was the US support that sustained Afghan insurgency. At this time the logic was that removing the pro-Soviet regime from power would be an important contribution to the peace efforts. However, this kind of support was dangerous. Once you support the religious fundamentalism it would be likely to result in a never-ending war against the non-Islamic forces. The guerrillas were armed with weapons, which were used against the Soviets. No one in the western circles, including the US objected when they declared their activities as a *jehad*. After the exit of the Soviet army, the guerrillas turned their guns against the US. In the post-Soviet context the battle with Taliban was not simple. The joint US-UK operations to remove this regime brought destruction on a very large scale. Their indiscriminate carpet-bombing[32] shook the world conscience. But it could not change the situation. The battle also led to the lack of confidence in the efficacy of the United Nations in resolving such vexed problems by peaceful methods.

The ethno-lingual composition of the population of Afghanistan is an example of ethnic diversity.[33] The people of Afghanistan speak a variety of languages. This is yet another indicator of the intermingling of different ethnic and ethno-lingual groups. Besides Pushto which is the predominant language, people converse in Dari (a variety of spoken *Farsi* or Persian). There are several Turkik dialects, such as Uzbek, Turkman, and the Tajik Persian, which are the common media of communication. It is understandable that the minority groups are concentrated in the regions adjacent to the Central Asian and the Iranian borders. Most of the population is Musalman[34] and among them Sunnis are predominant. On the other hand, an ethnic group called Hazaras, are by and large Shiites. The religion of the people in general is Islam with the only difference that it has a strong element of the mystic (*Sufi*) tradition. Religious leaders of the mystic tradition, known as *peers*, have an authoritarian hold on the minds of the people. English continues as a medium of education at the higher level as well as at the research level. It may be noted that the Hindus and Iranians have their colonies in all the major cities of Afghanistan because of their involvement in trade of dry fruits. Baluchi and Urdu are freely spoken in trading centres, or market places, depending, of course, on the location of the market place.

There is a controversy about the total number of ethnic groups in Afghanistan at the time of the Soviet withdrawal. Pushtuns claim that they are in majority. Their population is estimated as 7 million. Rough estimates indicate that the population is really around 24 million. Moreover, more than a million Pushtuns are still in Pakistan and in Iran as refugees. Those eventful years were marked by inter-ethnic conflict. This means that the ethnic diversity is the root cause of civil strife. The Pushtuns are mainly concentrated in the Kabul river valley, the Helmand river valley, and the Oasis of Qandahar. Their population has shifted to new areas in northern Afghanistan, such as Kunduz. There is a substantive proportion of Pushtun nomads and these nomads and the Hazaras have always quarrelled among themselves for pastures.

It is necessary to understand the ethnic composition of population. Pushtuns have been living mainly in mountainous parts of Afghanistan. They are also found all along the border with Pakistan. As one moves towards the west to the *registan* region, their number diminishes almost to zero. The main zone of concentration of the Hazaras is the central Hindu-Koh Range, while the Tajiks live in the eastern flank of the mountain range, people of Turkik origin, perhaps descendants of the Mongol or Turkik hordes are settled in northern parts of the country. The Persian Safavids occupied the area in sixteenth and seventeenth centuries A.C. They adopted a variety of Shiaism known as Twelve Shiaism, a variant of Shia Islam. The Hazaras have been marginalised politically and economically. They are mostly considered as the lower sections of the population and are engaged in menial jobs.

Turkmans and the Uzbeks are found in the northwestern part of the Hindu-Koh Range. More specifically, they live in the hills of Badghis. The rain-fed agriculture of food grains, like wheat, is their economic mainstay. They are the original Turkik tribes who came from central Asia. Their status is that of refugees. Thousands of Uzbeks fled away from their original homeland to avoid confrontation with the Soviet forces. Ismailis, a Shia group, were confined to the region north and northwest of the Salang pass. It was through this pass that they reached Mazar-e-Sharif. Ismailis live in small colonies in the midst of the Hazaras and Tajiks. Baluchis are the denizens of the rocky desert near the borders of Iran and Pakistan. Brahuis live in the southern and southwestern parts of Afghanistan. Agriculture and animal husbandry is their economic mainstay. Some of them are tenant farmers. Nuristanis live in isolated river valleys south of Badakhshan. The people in eastern Afghanistan embraced Islam in the late nineteenth century. Originally they were known as Kafiristanis. Farsiwan live in parts of Afghanistan bordering Iran. Qizilbashis are also known as Imami Shias. Another ethnic group known as Aimaqs have Turkik origin. They are concentrated in the western part of the Hindu-Koh range, west of the Hazarajat.

Afghanistan is a case of multi-ethnic composition of population. Ethnic groups, such as Baluchis, Pushtuns, Tajiks, Turkmans and Uzbeks are mostly Sunni and Hanafi. Shias have two groups—Imami Shias and the Ismaili Shias. Besides the population of the Muslims there were 20,000 Hindus and 10,000 Sikhs in the cities of Afghanistan as traders, merchants and moneylenders. They were the major players in the money market in the troublesome period 1992-96. They had left Afghanistan or migrated to safer areas in Afghanistan, such as Jalalabad. Kabul has a Christian Church serving the expatriate community. Afghan Christians are very few in numbers. It may be noted that Muslim conversion to Christianity is a difficult proposition.

Pushtuns, Baluchis, Sindhis and Punjabis Culture Realm

The population of Pakistan consists of Pushtuns, Baluchis, Sindhis and the Punjabis. They are the major ethnic groups. Whether they can be described as nationalities is a disputable issue. However, they have passed through several phases of quasi-feudalism before becoming the citizens of the state of Pakistan. The Punjabi identity is not exclusive to Pakistan. There are Punjabis on this side of the Indo-Pakistan border. So is the case with the Sindhis. Likewise, the Baluchis are concentrated along the international boundary between Pakistan and Iran. A substantive proportion of the Sindhi population migrated to India after the partition of India. They are living mostly in the states of Gujarat and Rajasthan.

Pakistan spreads over an area of 880,254 sq. km. and sustains a population of more than 156.5 million. The average density is 178 persons per sq. km. It is a country of great variety both in terms of natural environment as well as in the ethno-lingual composition of its population. This diversity is partly due to its geographical location. Several racial stocks came to this area in different phases in its history. Speaking in historical terms the different ethnic groups represented here include Dravidian, Indo-Aryan, Indo-Scythiyans and the Huns, including white Huns. But before these incursions, Dravidians were dominating the ethno-lingual scene. The latter incursions of the Indo-Aryans displaced the Dravidians who moved out of this region and shifted southwards in the Indian peninsula. Their remnants can be seen in Baluchistan, and the north Indian plain particularly, in the Jharkhand region of Bihar (now a separate state). Pakistan offered ideal conditions for racial intermixing. In some parts of Pakistan, Mongoloid stock is still traceable. It appears that the Greeks who came with Alexander left their imprints on the people of the subcontinent. Thus the diversity was further compounded as a result of centuries of racial inter-mixing. Lying astride the main routes of incursions and/or invasions, Pakistan has inherited a high degree of ethnic diversity.

There are several aspects of the ethnic composition, which deserve special mention. About 97 per cent of the population consists of Musalmans. They are mostly Sunnis, but Shiites also have a small concentration in certain parts of the country. There are other religious groups, such as the Hindus, Christians, Sikhs, Parsis (Zoroastrians) and the Buddhists. Even though the Constitution of Pakistan grants freedom of religion, the recent history shows that inter-linguistic relations are far from peaceful co-existence. Urdu, an Indic branch of the Indo-Iranian family is used widely in writings, particularly in literary works as well as print media. However, its official use notwithstanding, people claim Punjabi, Sindhi, Pushto, Baluchi and Brahmini as their first language. But Urdu is also a link language. English continues to be the language for inter-provincial and inter-national communication as well as higher education and research. Urdu has been ignored because of in-built constraints, and English has failed to attract the attention of the administrators in and outside the government. Reforms were necessary to put the economy on the fast track. Thus the import policy of the state has reduced the chances of survival of local industry. It amounts to saying that the policy has damaged the local entrepreneurship. Globalisation has further augmented the import of foreign goods. In fact, the US and other foreign countries find Pakistan as a potential buyer of their goods.

Bengali Culture Realm

Bengalis constitute another major ethno-lingual group. It is well-known history that the Bengali language and literature have unified east and west Bengal. In this context one may believe that the religious identities cannot bridge the gap between the countries and the people. On the Indian side, the state of West Bengal is the acumen of Bengali language, literature and culture. The state suffers from the floods as well as water-logging caused by them. The city of Kolkata (Calcutta) is the cultural centre of the Bengali language. The southern parts of the state have identical problems as faced by Bangladesh. The northern parts closer to the Indo-Nepalese border are prone to immigration from Nepal, mostly Gorkhas. The old demand for a state of Gorkha land has often led to civil strife. It is difficult to solve this problem. However, in recent past the Gorkhas have been given some powers to run their own affairs.

The political history of Bangladesh was written on two natural elements, water and mud brought down by the Ganga-Brahmaputra rivers and deposited in the deltaic tract, basically unstable due to the recurrent floods and branching of river courses. Bangladesh has a coastline of about 580 km. It forms a common border with India in the west as well as in the east and the north. The highlands in the east define the boundary with Tripura on the Indian side and Chittagong on the Bangladesh side. Geographers define the country's climate as tropical monsoon type. It has all the features of the Asiatic monsoon regime: hot and humid. Water-logging is very common because of the flat nature of the land and the recurrent changes in the river courses. Speaking in physical terms, there is too little land to accommodate its 129 million people. The burgeoning population has a demographic weight, not easily bearable by the land. Interestingly, the paucity of land is not a discouraging factor to population growth. Bangladeshis have a water-based economy—wet crops, such as paddy and jute are grown extensively, and fishing in rivers and the sea constitute a significant sector of production. Mighty streams—the Ganga, the Brahmaputra (locally known as Jumna) and their branches flow through the country. The mainstream of the Ganga beyond Farraka dam flows into Bangladesh as Padma. To the east of Pabna, near Gulando, the Padma receives the waters of Jumna. The combined stream flows retaining its name as Padma, till it receives Meghna to the north of Chandpur. The Meghna flows into the sea, dividing itself into numerous distributaries. Frequent floods cause dislocation of the peasant agriculture, loss of crops followed by starvation. The ancestors of Bangladeshis have lived with this syndrome for thousands of years. It is suggested that floods are not always negative; they also deposit a thin layer of silt to add to the natural fertility of the soil. But those who die due to floods or survive the floods, but die due to epidemics cannot share the imagined bounty.

Bangladesh is situated in the deltaic tract. The sea front of Bangladesh is prone to the hazard of the tropical cyclones, which visit the region on a regular basis during the southwest monsoon period (June to September). The positive side of the story is that these cyclones are rain bearing. These rain storms sustain the agricultural economy of Bangladesh. The country faces the challenge of storms and the devastation that follows. The cyclone, which hit the country in November 1970, left 300,000 people dead. Many more were rendered homeless. The cyclone caused widespread devastation and contributed to the growing dissatisfaction of the people. The deteriorating relations resulted in civil strife within the country and strained relations between the east and the west wings of Pakistan. The relief work in areas affected by such natural hazards is always difficult and leaves much to be desired. The people had no other option but to blame the Pakistani establishment for the loss of life and property. However, the natural hazards are not planned by the political establishments. The cyclone paved the way for the final breakdown of the political hold of Pakistan. The cyclone also made it necessary for India to intervene. The ouster of the Pakistani establishment helped India go ahead with a comprehensive relief programme and thus earn the goodwill of the Bangladeshi people. People were happy at this dramatic turn of happenings, a prelude to their liberation from the Pakistani establishment. The political situation changed fast and it was natural for India to celebrate the fall of the Pakistani army.

As stated earlier Bangladesh is a low-lying country. It is prone to flooding and the epidemics that follow. They result in mass hunger and death. This statement has serious implications. The human population as well as the population of domesticated animals is affected by the natural hazards. The two mighty streams of the subcontinent—Ganga and the Brahmaputra—pour their waters into the country before finally delivering them to the Bay of Bengal. The geographical location of Bangladesh appears to be a curse. Perhaps the impression one gets is that the new nation has been elbowed out and confined to a corner of the subcontinent—the most wet and the marshy corner. Climatic conditions apart, people have been reeling throughout the history—the history of cyclones, rains, floods, and the tidal waves. The news emanating from Bangladesh mostly report death and the devastation caused by the natural elements. About 95 per cent of the population lives in the country's sixty-five thousand villages. Towns are too few and farther apart. Dhaka, Chittagong, Khulna, Rajshahi and Narayanganj are the most well known urban places. These centres serve as nuclei for administration as well as for the collection of agricultural produce. This low proportion of urban population is due to the fact that the urban centres do not have much to offer by way of employment. The rural villages are dotted along the banks of the major rivers. The teeming millions dependent on agriculture and the allied activities lead an uncertain life. The old saying that Indian agriculture is a gamble with monsoon holds good even today.

Earlier known as East Bengal, now Bangladesh is the fortress of the Bengali language par excellence. The Indian state of West Bengal, of course, is a part of the same cultural realm. The literature in the Bengali language binds together the people of the West and the East Bengal. It is this link that can be manipulated to create the chasm between the Bengalis on this side of the international border and those on the other side because of the difference in the religious identity. Very often the Bengalis who have been living in the large metropolitan centres of India, such as Delhi and Mumbai are treated as illegal infiltrators. Languages spoken other than Bengali include Urdu, and a variety of dialects of the Tibeto-Burmese family. Hindus constitute a significant proportion of the population of Bangladesh. However, their way of life and the spoken language is so similar that it is easy to mistake them for the infiltrators. In phases when religion became the basis of the partition of India, religious identity became a source of rioting and mass killings. Both India and Pakistan took advantage of this situation for further strengthening the divide. The tribal communities, such as the Chakmas, Moghs, Santhals and the Garos, are concentrated in the regions bordering Assam and Meghalaya, and in the Chittagong Hills District on the border of Tripura. These ethno-lingual groups in Tripura have little scope to protect themselves from the fast growing population of non-tribal Bengali peasantry. These ethnic groups along with the Buddhists and Christians are the main minority groups.

While it is true that making generalisations on the inter-community relations may be risky, the ground reality is that underdevelopment and the fast growth of population are the major problems of the Bangladeshi nation.

All said and done, the state has not played its role as a catalytic agent. This may be due to the over-arching problems of political instability. The recurrent imposition of the martial law and the dictatorship of the top brass of military seemed to be the only alternative for sustaining the nation.

Nepali/Newari Culture Realm

The population of Nepal consists of two major ethnic groups, Indo-Nepalese and Tibeto-Mongoloids. Racial inter-mixing has led to further diversification. The Indo-Nepalese groups came from the south and the Tibeto-Nepalese groups came from the north. Ethno-linguists recognise that the Himalayan kingdom presents a good example of hybridisation of racial stocks and linguistic traits. The mainstream of the Tibeto-Nepalese ethnic groups consists of the Sherpas, known as the world's most skilled mountaineers, the Gurungs and the Gorkhas. The Gurungs are traditionally soldiers. One view is that the Gorkhas have been derived from the intermixing of different ethnic groups living in Nepal. The population of the kingdom is of mixed origin. A major concentration of population is seen in the Kathmandu valley and an equally important cluster in the *terai*, where the densities are high because of the rich agricultural base. Understandably, the northern sub-Himalayan regions are sparsely populated and the population is mostly rural. The urban nodes are few and farther apart: Kathmandu, Bairatnagar and Lalitpur. The relations between India and Nepal are centuries old. Nepal's bi-lateral relations with India and China have always been cordial. However, the cultural influence of India is more visible as compared to China. The official language is the Nepali, which belongs to the Indo-Aryan family and is close to Hindi. About one-half of the Nepalese population speaks Nepali. There are as many as thirty other languages, which are spoken in daily communication. Notable among these languages is Newari, which is quite popular as a spoken language. The population of the Himalayan kingdom is largely Hindu (85 per cent). This makes Nepal the only Hindu country in the world, other than India. Buddhists account for 8 per cent of the population and Muslim population is just 4 per cent. However, Nepalese Hinduism assimilated elements of Buddhism. It was also influenced by the Himalayan cults. Even Islam has a Nepalese version of its own with strong Buddhist underpinnings.

Bhutanese Culture Realm

The hill economy of Bhutan is essentially based on agriculture and processing of the raw materials derived from agricultural and pastoral produce. Paddy, corn and potatoes are the main crops grown on the terraced slopes. Fruits, such as oranges, apples, pears and plums are exported in raw form as well as in processed form, like jams and marmalades. Thimphu, the capital city of the kingdom, is a small town. Bhutanese follow a Buddhist traditional religion, *i.e.*, Lamaistic form of Mahayana Buddhism. The Bhutanese speak Tibeto-Burmese dialect, known as Dzongkha. It is believed that it was introduced in Bhutan in the mid-sixteenth century. This is also the official language of the kingdom.

Not much information is available on the history of Bhutan. However, historians cite evidence to show that the territory was under the rule of Indian princes. Eventually, the Bhutias drove them out. Bhutan passed through a period of struggle for power for more than almost three centuries (1300-1600 A.C.).

Sinhalese Culture Realm

An overwhelming proportion of the population of Sri Lanka consists of the Sinhalese. They comprise three-fourths of the island population. Tamils are the largest minority group. The Tamils of Sri Lanka are divided into two groups—Sinhalese Tamils and Indian Tamils, together accounting for 18 per cent of the population of the island. Other minority groups include the Moors, Burghers, Malays and the Veddas. The Arabs came as traders, boatmen and fishermen. The Burghers are the descendants of the old Dutch and British settlers. The Veddas is an aboriginal tribal group. Malays and other groups migrated from India to Sri Lanka in the hoary past. Sri Lanka is not entirely located to the south of the Indian peninsula. It lies rather to the east of the state of Tamil Nadu. Tamil migration was a natural corollary of the geographical location of the island. The recent conflict between the Sinhalese and the Tamils is unfortunate. It has resulted in widespread devastation and death. The Norwegian diplomatic efforts to bring them together are laudable by all means. In such situations talks around a table may be more productive than exchange of bullets. There are distinctive differences between the Sinhalese Tamils and the Indian Tamils. This distinction depends on the time of their arrival. The Sinhalese Tamils obviously came earlier while the Indian Tamils are recent arrivals. The ethnic differences between them are further widened when it is discovered that they belong to the two different language families. Tamil belongs to the Dravidian family, while Sinhalese is an Indo-Aryan speech form. Sinhalese is spoken by about 70 per cent of the population of the island nation. English is widely spoken. English language continues to be the medium of communication and used in all administrative work. It is also the language of education at the higher level and research.

Sri Lanka has a population of about 20 million. The island nation encompasses a geographical area of 66,000 sq. km. The island is characterised by vast diversities in terms of terrain, climate, soils and the natural resources with a substantial scope for further development. Students of cultural history have noted Sri Lankan Buddhism as a category, perhaps a blend of Hinduism and Islam. The island population consists of Buddhists (69 per cent), Hindus (15 per cent), Christians (8 per cent) and Muslims (8 per cent). The Veddas, supposedly an aboriginal group, are also found in the isolated parts of Sri Lanka. But their population is not significant. Most of the population is rural and dependent on agriculture and allied activities. Urbanisation is slow. A high foreign exchange-yielding plantation-based economy and agriculture generate work for the lowly and the downtrodden sections of population. The economy has not so far been able to absorb them and put them on the path of development and prosperity. The plantation crops are exported, but the back flow of the export earnings is by and large small. The income generated is, by and large, impounded by the planters, which is understandable. But the quality of life of the daily-wage workers has not improved. Plantation agriculture has little role to play in the upliftment of the toiling masses, even though it has the potential to do so. Unfortunately, the surplus generated through export trade has, by and large, not been properly used to help the poor.

The island nation has a historical record spanning over at least a period of twenty-six centuries, from 504 B.C. onwards. Not much information is available on the history of the island from sixth century B.C. to sixth century A.C. *Mahavamsa*, local chronicle, maintained by the Buddhist Bhikshus is believed to be the most important source of historical information. It contains information for a period of around 1200 years from sixth century A.C. to nineteenth century A.C. The salient features of the political history may be recapitulated. Vijaya, the legendary prince, is credited for conquering the island in c. 405 B.C. Basham has suggested that the Hindu prince, Vijaya, came from the northeastern part of India. Apparently, it appears like a legend and historians have tried to reconstruct the missing links in the Vijaya story. It is widely believed that the prince established his authority by subjugating the local aboriginal groups known as the Veddas. Prince Vijaya married a local woman to establish his credibility.

His *modus operandi* was simple. He married a local woman; encouraged emigration from peninsular India; declared himself as the undisputed ruler of the entire island and named the country as Sinhala. Prince Vijaya's successors continued to rule over the island for several centuries. They adopted Buddhism as the state religion in the third century B.C. and initiated several activities to promote the Sinhala civilisation. The Sinhala dynasty was instrumental in building large irrigation projects, which reveals their engineering skills. The kingdom also introduced a new architectural style and took steps to restore many shrines, locally called *dasobas*. The island was taken over by an enterprising Tamil king. The Tamil control lasted for nine centuries from the late third century to twelfth century

A.C. The religious history of the island underwent a major change. It may be recalled that the Chinese forces ruled the island for a period of thirty years. Like India, Sinhala also passed through a phase when the island was divided into petty kingdoms. These little kingdoms disappeared before the Dutch annexed the island to their vast empire in Southeast Asia. The subsequent history of the island was related to the competition between the maritime European powers like the Dutch and the Portuguese. The Portuguese were great seafarers and traders. They established a fort and a trading port at Colombo in the sixteenth century A.C. These developments further promoted their overseas trade. By the end of the sixteenth century, the Portuguese hold on the island was almost complete. However, the Dutch challenged their supremacy and launched a series of attacks (1638-1639 A.C.) to remove the Portuguese stranglehold on the island. The Dutch-Portuguese rivalry came to an end in 1658, when the Dutch established their hold over most of the island territory. The Kandy kingdom was, however, spared. After establishing themselves in India, the British dispatched a force to the island in 1795. They lost no time in annexing the whole island. Three years later in 1798, the entire island excepting the Kandy kingdom became a crown colony of the British. According to a treaty signed in 1802, the island was annexed to the British Empire. As a consequence, the entire island was formally ceded to Great Britain. The British lost no time in occupying the Kandy kingdom, which was declared a crown colony in 1815. The people did not accept the British ascendancy and several rebellions took place during 1817 and 1848. However, the struggle for freedom lasted longer than imagined. It took at least a hundred years in achieving this goal. Full freedom came in February 1948. Under the British, the island passed through a phase of tripartite conflict between the Buddhists Sinhalese peasants, the Muslim traders, moneylenders and the British. The struggle for freedom continued unabated over all these years.

Hindi/Urdu Culture Realm

The spoken forms of Hindi/Urdu serve as languages of inter-communication over vast regions of the subcontinent. In fact, they are the lingua franca of the people, particularly in north India. The colloquial form of these languages may vary from region to region. Urdu, for example, has several spoken forms, such as Lucknowi, Deccani and the Punjabi. The regional languages have influenced Urdu, particularly in its colloquial forms. Hindi, on the other hand, has quite a number of dialects, some of them having an independent status so much so that they are often recognised as separate languages. This is particularly true for Braj Bhaka, Awadhi, Maithili, Maghadhi (Maghai), and Marwari, to mention only a few. They have a well-recognised literary tradition spanning hundreds of years. Urdu has the distinction of being the lingua franca of people from Kabul and Herat both in Afghanistan, and Kalat in Baluchistan in the west-northwest to Guwahati and Manipur in the east-northeast and from Baltistan, Kargil and Ladakh in the north to Madurai and Kanyakumari in the south. In the large metropolitan cities, such as Mumbai, Kolkata and Delhi the population consists of different language groups.

Hindi/Urdu serve as the language of communication in the market place. They may be described as pigeon languages. They have assimilated the loanwords, phrases and phonological elements from different regional languages. The loanwords are transformed to such an extent that they lose their original character. It may be noted that the spoken form of the two languages is similar. Problems arise when they are written in two different scripts: Hindi in the Devanagari script and Urdu in the Arabic script. Interestingly, Sindhi, Punjabi and Kashmiri are also written in the Arabic script with slight differences. Punjabi, particularly in the Indian state of Punjab has adopted Gurmukhi as its script. Punjabi in Pakistan, on the other hand, has retained its original Arabic script. Urdu has also served as the medium of instruction in the northwestern parts of the subcontinent, like Kashmir, Punjab and Sind. The old generation of the Punjabis and the Sindhis in India read Urdu newspapers. However, all these languages have a long literary tradition. These generalisations may be disputable. The origins of controversy lie in politics and not in linguistics. There are two main factors, which contributed to the controversy. First, the rising tide of communalism in India before independence tried to sidetrack Urdu as a language of the Muslims and Hindi as the language of the Hindus of north India from Uttar Pradesh and Bihar to Madhya Pradesh and Rajasthan. The same logic was applied in India after partition that with Muslims who migrated to Pakistan Urdu had no place in India. The Indian National Congress, particularly its secular minded leaders, such as Gandhi and Nehru opposed this vehemently. Nehru during his tenure of seventeen years left no stone unturned in ensuring a secure place for Urdu in India. Institutions, such as Anjuman Tarraqui-e-Urdu were given financial support to continue their activities. All India Radio continued to telecast its Urdu programmes. Urdu Academies were established in all states of the Indian Union.

Hindi/Urdu in their role as lingua franca served two purposes: (a) when people with different ethno-lingual background came together they communicated in Hindi/Urdu; and (b) Urdu was adopted as a medium of instruction and of literary expression in north and northwestern India. Urdu also had the distinction of serving as a medium of literary expression in other language regions such as Orissa, West Bengal, Maharashtra, Andhra Pradesh, Karnataka and Tamil Nadu to mention only a few. Urdu Academies in these states further augmented the progress of Urdu. Similarly, Hindi Academies were established in all states of the Indian union. These developments have resulted in significant regional variations in the written form of Urdu, which developed nuances of its own in different regions.

The adoption of two different scripts for Hindi and Urdu gradually widened the gap between the two. For instance, Hindi borrowed words from Sanskrit and regional *Prakrits* while Urdu leaned more towards Persian and Arabic. However, like standard Urdu, standard Hindi is a form of *Khari Boli*. Regional dialects of Hindi are different. For example, words and sounds are rounded; r becomes l or it is dropped altogether. If you want to say *kar diya* in Khari Boli, in Braj Bhaka, it will assume the following form: *kaddiyo*. The give-and-take over centuries of shared history between the standard form of Hindi and its numerous regional dialects have resulted in the development of regional variants. The spoken languages were variantly described as Hindi, Hindvi, Urdu or Hindustani. These variations in name do not matter much because the lingua franca is the same. It would not be out of place if the origins of Hindi-Urdu are recollected. The very word Urdu came into being as the original *lashkari* dialect, in other words, the language of the army. The Urdu bazaar near the Jama Masjid of Delhi was essentially a *lashkari* bazaar, meaning thereby a market place for the army. The open ground between the Jama Masjid and the Red fort in Delhi was used for camping by the army. Urdu's literary tradition eventually connected it to the official language used in the courts of kings and emperors who ruled over the parts or the whole of the subcontinent. The time period roughly may be mentioned as the beginning of the first millennium A.C. and onwards. It is noteworthy that Urdu was never recognised as the court language. Many Urdu poets including Mir and Ghalib called Urdu as *Rekhti*, which literally means a second grade language not to be used by the elites. Mir wrote his autobiography in Persian, even though his poetry was composed in Urdu. Perhaps Ghalib was the first Urdu poet of standing who wrote letters to his friends in Urdu. In fact, Ghalib is recognised as a stylist of Urdu prose and converted the written language into a dialogue. With reference to *rekhti* the following lines from Ghalib may be quoted:

Rekhti ke tum hi ustad nahin ho Ghalib
Kehte hein agle zamane main koi Mir bhi tha
(O Ghalib you are not the only master of *rekhti*, I am told Mir was also a pioneer, who composed his ghazals in Urdu.)

The language of the Urdu ghazal was characterised by its numerous loanwords drawn from many languages. It was also supposed to be used mainly by women in both written and spoken forms in an unpersianised style. It appears that Urdu originated in the early medieval period when Turks (Turksmans), Uzbeks, Parsis, Mongols/Mughals came to the subcontinent and established their dynastic rule. The spoken dialect was a mix of many languages. Hindi/Urdu became the lingua franca of the people who came from different ethno-lingual background. Persian was used as a written language as late as Ghalib. In his own view, Ghalib insisted that his main contribution was his *ghazals* written in the Persian language and not in Urdu. In fact, he described his Urdu *Dewan* (collection of *ghazals*) as *berang*. *Biguzar az majmua-e-Urdu ke berang-e man ast* (read my collection of Urdu *ghazal* rather superficially, you won't enjoy it). Paradoxically, today Ghalib's fame is due to his Urdu collection of *ghazals*. In fact, it is his Urdu *ghazal* that immortalised Ghalib.

It may be concluded that Hindi/Urdu was adopted as a language of communication in two main forms: (a) as a spoken language in personal inter-communication; and (b) as a written literary language. Even today it is spoken in the market place as already highlighted earlier. In its written form Urdu was adopted by writers in areas, such as Punjab, Kashmir, Sind, Baluchistan, Rajasthan and other parts of the subcontinent for their literary expression. In northern India from Delhi to Bengal, Madhya Pradesh and Maharashtra, Urdu is the medium of oral communication, along with the local and regional dialects. Till recently Urdu was the first choice for all literary work. This difference in the use of language is remarkable. We know that the regional languages in Pakistan: Baluchi, Sindhi, Punjabi, etc., are used by and large for oral communication. Seen in this background Hindi/Urdu cuts across the linguistic boundaries within the subcontinent. However, in the spoken form its realm, or language domain, can be described as trans-regional, even subcontinental.

Understandably, people from different ethnic backgrounds were recruited for the army and sooner or later a colloquial language emerged for communication between them and among themselves. This spoken language came to be known as Urdu. The inter-communication between the different contingents of the army initiated a process of intermixing. Eventually, this resulted in the emergence of a common dialect with loanwords borrowed from Turkish, more specifically Uzbek, Arabic, Persian, Dari, Sanskrit, regional *Prakrits*, Greek, English and so on. For example, lantern became *lalten*, ticket became *tikat*, pantaloon became *patloon*, and orderly was expressed as *ardali*. The Urdu word *maiz* was derived from the Portugese.

Dravidian Culture Realm

The story of the Dravidian culture realm is recapitulating the glory of the South India, its culture and civilisation. Technically, the boundary between the North and South is the boundary of the Indo-Aryan North and the Dravidian South. It is noteworthy that the boundary is not parallel to any latitude. In fact, it is a diagonal line joining Goa with Raigarh in Chhattisgarh. South of this line lie the domains of the Dravidian family of languages. It is generally believed that Dravidian is as old, or even older than the Harappan culture. At one time in the history of linguistic evolution of the subcontinent, there was a stage when Dravidian languages occupied most of the northern parts of the subcontinent from Baluchistan to Chhotanagpur plateau. A drastic change in language came with the advent of Indo-Aryans in the subcontinent sometime in the middle of the second millennium B.C. Initially, the Indo-Aryans were confined to the Saptasindhva region in northwestrn India. However, the Indo-Aryans started moving towards the east and went up to the Rajmahal hills. They further moved to the southeast and the south of the Vindhyan mountain range. Tamil and the other languages of the Dravidian family moved towards the southern peninsular region. With this shift to the peninsular region of the Dravidians, a nucleus of Tamil language and culture emerged in the Kaveri valley. While the Dravidians migrated towards the plateau region, they left behind the pockets of Dravidian languages in several parts of the subcontinent. The dialects, such as Bruhi in Baluchistan, Kuruk Oraon, the Malto in Jharkhand and the Gondi in parts of Chhattisgarh and Andhra Pradesh were the remnants of that Dravidian past. Huen Tsaung who visited the subcontinent in the early part of seventh century A.C. has mentioned the names of the south Indian kingdoms, like Chola, Pandya, Chera and Madura. Historical text refers to the Chola, Pandya,

Bellals and Telengana kingdoms. The northern border of these kingdoms is roughly parallel to the Godavari river. The Telengana kingdom was situated on the west and southwest of Godavari river in the peninsula. The Cholas, the Madura and Pandya kingdoms lay to the south of the Krishna and its tributaries Tungabhadra. The region from Kanyakumari to Kaveri was the hub of Pandya kingdom. In the complex of hills of the Western Ghats lay the kingdom of the Bellals. The picture becomes clear by the early thirteenth century. By the end of the fourteenth century the kings of Vijayanagar were controlling all the regions south of the Krishna and Tungabhadra. The Afghan kings of the early fourteenth century expanded their hegemony over most of the peninsular region sparing only the Madura kingdom. Madura kingdom was confined to the region south of the Vaigai river. The Madura kings were also forced to pay the tribute to the Afghan kings of the north India. The Vijayanagar kings were holding the fort up to the end of sixteenth century A.C. Southern kingdoms, such as Golkonda, Bijapur and Polygars were in control of the peninsular Dravidian kingdoms. The political map of south India in 1751, 45 years after the death of Aurangzeb, shows that the Dravidian south was in control of Nizam's dominions up to Krishna and Tungabhadra. The eastern region in the Eastern Ghats and the littoral region were occupied by the Carnatic Circars, the interior of the peninsular plateau was in the possession of Mysore. The Malabar coast south of Goa was divided among the Polygars in the north and the state of Kochin and Travancore in the south.

By the middle of the eighth century A.C. a king named Rajaraja the great (985-1014 A.C.) defeated the Pallavas of Kanchi. He also conquered the Vengi kingdom. He reduced the Pallavas of Kanchi who were ruling this region for almost eight centuries. He also subjugated the Sri Lankan kings and became a power to reckon with all over south India.

The earliest Tamil texts show that the real founder of Tamil language and grammar was no other than Agastya. There is a controversy on the origins of Tamils. It is argued that Tolkappiyam was the most celebrated of the twelve pupils of Agastya, the great sage. It is argued further that the original grammar of the Tamil language was Agattiyam. According to the tradition, Agastya's work was a source of Tolkappiyam. Historians believe that there is no mention of Agastya in the Tolkappiyam; even Panambaranar who wrote the preface to the Tolkappiyam does not mention him.

The Changing Linguistic Map of the Subcontinent

The data on language and dialect furnish the most important evidence of the socio-cultural diversity of the subcontinent. It is noteworthy that this diversity makes the subcontinent a babble of 'tongues'. However, this diversity does not mean mutual exclusivity. Language groups have always interacted with the neighbouring groups implying a long process of inter-linguistic and inter-regional communication. The linguistic give-and-take has continued for centuries. In fact, the reality is that a continuous exchange of words, phrases and colloquial structures has always been going on through the ages. This has resulted in building the bridges between the spoken languages. The best example of this mixture is seen in the case of Hindi/Urdu. The subcontinent does not present a picture of linguistic pluralism. A spoken language serves as a bridge, not a cleavage. The language and the dialect of human communication cannot be supposed to exist in watertight compartments. They may be born in isolation, but sooner or later they spread their area and this results in contact with other linguistic groups.

The inter-communication between one language and the other always becomes a necessary step towards change along the contact zone. The history of languages shows that different ethnic communities lived together in a shared territory. However, it may be stated at the very outset that delimitation of language domains is not an easy exercise. Languages merge with one another almost imperceptibly. Languages have interacted with one another and expanded over contiguous geographical space. Thus change in language does not come abruptly. There is always a transitional zone through which one language gives way to the other. Many a time in history small language groups have been assimilated into the larger ones. Grierson's linguistic survey of India revealed that there were as many as 179 languages and 544 dialects. The survey covered all regions of the subcontinent including Pakistan, Nepal, Bangladesh, and Sri Lanka. Unfortunately, Grierson's data is, to some extent, misleading. It appears that he was confused on the question of classification of languages and dialects. It does not diminish the value of his work, which was pioneering and original. Taxonomically, the dialects are often a variant of the same language. Moreover, of the 179 languages listed in the survey, as many as 116 were spoken by minor tribal groups. They belong to the Sino-Tibetan family. After independence, language data was most systematically collected during the 1961 census of India. The census recorded as many as 187 languages. The data was subjected to a classification scheme under the supervision of a competent linguist attached to the census department. Of the 187 languages, 94 were spoken by small ethnic groups of less than 10,000 persons. It is noteworthy that as a whole an overwhelming majority—constituting 97 per cent of the total population covered, were affiliated to 23 major languages. Thus the diversity is contributed by 3 per cent of the population. The ground reality is that census enumerators were bound to record the response of the people on mother tongue. People insisted that their mother tongue was 'x' or 'y' even though these categories were not existent. For example, many people insisted that their mother tongue was *Pathani* or *Sitapuri*. These mother tongues were invented and were not there in any scheme of taxonomic classification. Naturally, the data posed serious problems of interpretation.

The different language groups were redistributed over the adjoining areas. This is an example of dispersal. The language groups are also clustered and concentrated in the nuclear regions. This is an example of accretion. In this process of historical evolution language groups were dispersed in space. This reduced their proportion in the total population from census to census. Sometimes in this process of dispersal small languages were merged with larger ones. There were also cases of overlap and gradually some of the groups were assimilated into other languages and disappeared. An eminent Indian linguist, Khubchandani, noted that the population of the Indian subcontinent due to 'sustained interaction and common legacies have developed a common way to interpret, to share experiences to think'. Khubchandani defined this phenomenon as 'organic plurality'. In his view the basic element of the subcontinent's linguistic unity is observed in the fuzzy nature of linguistic boundaries. In fact, he described the phenomenon of shifting loyalty of the people in the declaration of their mother tongue. He describes the process as 'fluidity' in language identity. This is due to the fact that the same people responded that their mother tongue was 'x' and not 'y'. At a subsequent census the same people declared some other dialect as their mother tongue.

The situation in the subcontinent is a kind of complementarity of inter-group and intra-group communication. Historical data indicate that the Dravidian languages were the first major group which arrived in the subcontinent. They occupied almost the whole of the subcontinent. Much later in time, a new group appeared which is generally recognised as the speakers of the Indo-Aryan languages. Archaeologists have observed that they emerged on the scene not earlier than the middle of the second millennium B.C. With the displacement of the Dravidian people by the Indo-Aryans, Dravidian languages shifted to the peninsular region. The Indo-Aryan region also experienced significant changes in the spoken language. This change implied a transition from the Vedic Sanskritic languages to Asokan Prakrits like the Pali. The evolution of the Indo-Aryan languages must have taken place during the period from 200 B.C. to 200 A.C. It is noteworthy that the evolution also led to a specific geographical pattern of distribution and language zonation. There were several phases through which the linguistic transition occurred. These stages represent an evolutionary change often described as middle Indo-Aryan, late middle Indo-Aryan and new Indo-Aryan. The north Indian region passed through a rigorous process of language mixing. The spoken language came to be known as Hindi/Urdu, sometimes referred to as Hindustani. These speech forms emerged as a result of inter-communication between the different *apbhramsa* and the foreign languages, which came to the subcontinent in different phases of history. These languages with which an intense process of inter-communication started can be listed as Greek, Turkik, Arabic, Parthian or Persian and Dari. The following line of Amir Khusro reflects powerfully the difficulties of inter-communication with the foreign language groups:

Zuban-e-yar-e-man Turki wa man Turki namidanam
(The language of my beloved is Turkik, but
I do not know Turkik)

This was the dilemma of the people of north India during Khusro's lifetime (1211-1290 A.C.).

It is generally known that the linguistic map of the subcontinent has always been changing. This change is due to: (a) evolutionary process unfolding itself in geographical space, and (b) due to intermixing of ethnic groups with diverse origins as a result of migration and linguistic fusion. However, one can take stock of the situation by formulating a set of hypotheses.

First, when the people with different ethno-lingual contexts meet, a change in spoken language is imminent. For example, the mixing of population in the contact areas leads to a language change from one language group to the other. Secondly, exposure to long-range migration of ethnic and culture-groups leads to borrowing of words and phrases. The interactive process between the host and the guest languages necessitated this change. Thirdly, education even up to the primary level, in a regional language, which is not the mother tongue of the learners, is also a strong basis of change in language. In this case, the school provides the necessary environment for the language change, linguistic shift or bilingualism. Fourthly, the contact with the market forces and the infiltration of the outside trading communities is likely to be a major agent of change in language structure and phonology. Fifthly, the language change is related to the change in administration, reorganisation of states, occupational background of the people, place of residence of the out-groups, educational status and population change due to migration.

Early migrations in the history of the humankind resulted in the expansion or contraction of the language domains. People moved from one region to the other in the very process of peopling of a given territory. This means that early migrations left an impact on the language as well as culture of the people in host regions. Language changes in regions of departure, language also changes in the regions of arrival. By regions of departure we mean a geographical area, which is an out migrating zone and from where people move out for different reasons. On the other hand, regions of arrival mean a destination to which the people are heading for.

Development is also a major agent of language change. There are several factors, which regulate the progressive process of development. One such factor is urbanisation, which implies a shift in residence from the village to the town. It also implies a shift in residence from a town to a city or from a city to another city. Urbanisation also implies the expansion of cities, which eventually leads to the absorption of rural population into the boundaries of a city at a certain point of time. This change is explicitly a product of expanding boundaries of the city. This shows that the rural population has not migrated out; only the city has expanded and thus encroached upon the villages on its periphery. The best example of this process is seen in Delhi, the National Capital Territory of Indian Republic, which has assimilated hundreds of rural villages. Development also implies industrialisation. It plays an important role as a catalyst of development. It attracts labour, raw materials, and capital resources from the neighbouring, or distant areas. It is also related to the changes in the administrative set-up. For example, the area of jurisdiction of a municipal committee, or a municipal corporation expands. However, it cannot wipe out the identity, way of life, social organisation, ethnicity and language structure of a certain village, which has now been surrounded by the expanding city. The urban administrative authority cannot impose laws on the rural masses thus brought within the urban net. In the case of the subcontinent it is the tradition and not the law, which is paramount. In many cases law is an extension of tradition. The rural folk drawn within the ambit of the city are governed by their own tradition, social practice and custom rather than by the municipal bye-laws. In fact, municipal bye-laws are irrelevant to them. There is no municipal authority that can persuade the semi-urbanised rural folk to conform to the bye-laws.

The subcontinent passed through several phases in the history of industrialisation. The main phases can be identified as: pre-British, British and post-British. The pre-British phase was a continuation of the artisan industry of the medieval times, which had brought name and fame to the country, particularly the artisans engaged in this household industry sector. Initially, the artisan industry of the subcontinent was retained and supported by the East India Company. But soon the British realised that further continuation of the artisan goods industry and their export to other parts of the world would affect the upcoming British factory industry at home. Therefore, systematic steps were taken not to promote the artisan industry in India. Instead, the British induced the export of raw materials from India to Britain to support their factory industry. This would mean that India would be converted into an exporter of raw materials and an importer of the British factory industry goods. However, there were several sectors of the manufacture of industrial goods in which the British were forced to initiate the industrial activity within the subcontinent. The situation changed because of the First World War (1914-18), which disrupted the outflow of raw materials from the subcontinent and the inflow of factory goods into India in return. In fact, war imposed a new regime of the manufacture of goods needed by the armed forces. This appeared to be the best alternative at that time. There were other industries, such as cotton and jute textiles, sugar, *khandsari*, vegetable oil, and goods for the use of armed forces. These industries were located in different parts of the subcontinent and they were geared to meet the local demand.

The British policy on industrialisation in India changed drastically between the First and the Second World Wars. The indigenous industry of the subcontinent got an impetus to develop in the core sectors. India and Pakistan made a vast breakthrough in the post-British period in the field of industrialisation.

Selected Itineraries

In this section we have tried to chart out a set of itineraries in order to take stock of the ethno-lingual composition of population of the subcontinent.

From Kabul to Qandahar

A journey from Kabul to Qandahar will take one to the glory of rocks and the relatively barren hills lined up as if on display, denuded of their flora and much of their past history written on their face. But the wilderness of the landscape is lost in the tiny rural settlements of the people who are the denizens of the rocky land. This region is drained by the Helmand and its tributaries like Arghandab and the Tarnak rivers. These streams carry some water even in the lean season in their otherwise dry channels. The Tarnak meets the Helmand west of Qandahar. Who are the people whom one meets on the way? Pushtuns, they are there, in their traditional attire, tall, turbaned and handsome with their self-confidence writ large on their faces and the deep lines indicating the struggle for existence. They are the true Afghans, even though divided on the basis of clans (*jirga*) and sub-clans, Durrani and Ghilzai being the most prominent among them. The journey takes one to the two climatic regimes marking the transition from the semi-arid to the arid. The Tarnak river is virtually the boundary between the semi-arid region around Qandahar and the arid zone further south described as Registan. As one travels from Kabul to Qandahar, one notices small pockets of the Tajik settlements, where Tajik form of Persian is spoken instead of Pushto. Qandahar is known for its orchards growing fruit crops, such as pomegranates and the grapes. The carpet-bombing by the allied forces disrupted the supply of these fruits, both dry and green, hitting their plantation economy badly.

Yet another cluster of Tajik settlements lies to the south of Qandahar, before one reaches the settlements of Baluchis along the Pak-Afghan border north of Baluchistan.

From Kabul to Iranian Border

The Hazaras occupy a vast expanse of land west of Kabul in north-central Afghanistan from Firoz-Koh to Helmand. To the west of the Hazarajat lies the territory of the Aimaqs, interspersed with Hazaras, and Tajiks, north of Herat. Hazaras also have several pockets of their settlements in the region close to the Iranian border. Their settlements are interspersed with village clusters of the Aimaqs.

From Kabul to Herat

The road journey from Kabul to Herat passes through the difficult terrain on the southern and the southwestern slopes of the Hindu- Koh Range. The road runs parallel to the Harirud river up to Herat south of the Paropamisus range. The mountain range overlooks the road. The road passes through another cluster of the Hazara settlements beyond Chaghchareen where one meets the Aimaqs. Herat is situated on the north of the Tajik settlements and south of the enclaves of the Aimaqs. Turkmens, another ethnic group, live in the region north of Herat and across the Paropamisus range. They are also settled along the Iranian border and the border between Afghanistan and Turkmenistan. Turkmens are also located on the road from Kabul to Mazar-e-Sharif and beyond. The road runs through Chrikar and Baghban passes and descends down the heights of the Hindu-Koh Range.

From Kabul to Mazar-e-Sharif

From Kabul on way to Mazar-e-Sharif one passes through the clusters of villages of Pushtuns, Hazaras, and the Uzbeks. The Turkmens dominate the ethnic landscape north of Mazar-e-Sharif. The Turkmen settlements in this area have come into being as a result of their dispersal in geographical area. Nuristanis have their settlements in the region lying to the north and the northeast of Kabul. The road journey from Kabul to Lahore runs parallel to the Kabul river. Beyond the Kabul river valley one crosses the Khyber Pass to reach Bandikotal, Peshawar and Rawalpindi, all situated in Pakistan. Beyond Rawalpindi the road connects the cities of Gujar Khan, Jhelum, Gujarat, Gujranwala with Lahore. In this sector the road crosses the Indus near Nowshera, and then it crosses the Jhelum, Chenab and Ravi rivers before it reaches Lahore. A branch of this road connects Peshawar and Muzaffarabad, the capital of the occupied Kashmir. From Muzaffarabad the road crosses the Jhelum gorge to reach Sopore and Srinagar. However, while international boundary is crossed, the physical type of man does not change on either side of the Durand Line. Pushto remains the dialect of communication. It is sometimes replaced by Dari, a variant of Persian. From Peshawar to Muzáffarabad and on to Sopore, Kashmiri continues to be the lingua franca.

From Peshawar to Srinagar

Srinagar is the hub of the valley of Kashmir. The city has remained as the capital of the Kashmiri language and culture. However, Gurjari or Gujjari mixed with Urdu is the common dialect. Gujjars and the Gujjar Bakrwals are trans-humant groups who remain on the move all the year round. They have their own traditional treks. There are several places whose names are associated with Gujjars, like Gujar Khan, Gujarat and Gujranwala. One crosses the international boundary between India and Pakistan in the region to reach Amritsar. Punjabi is commonly spoken on either side of the international border. There are several language divides between the Pir Panjal (Pir Pantsal) Range and the Great Himalayan Range. These divides include a change from Dogri to Kashmiri languages, and between tribal dialects and Kashmiri, south of Gilgit. The Banihal pass route takes one to Jammu via Udhampur. Besides Dogri there are other dialects, which are frequently spoken. There are small tribal communities who are regarded as the original inhabitants. Punch, Riasi, Akhnoor and Udhampur are the centres of Dogri culture and dialect. East of Srinagar and across the Great axial range of the Himalayas one finds that the mountains have acted as a watershed between Kashmiri and the Kishtwari. South of Jammu, Punjabi and Dogri are the main speech forms. In Pathankot and Kathua, people communicate in variants of the Punjabi and Dogri dialects. The Sikhs are concentrated in this area, even though they use the Punjabi language in inter-personal communication. The language divide is sharply defined because of the mountainous terrain. East of the Ravi, one reaches the heartland of Punjabi. The spoken language changes from Lehnda to Seraiki, particularly in the region between Pathankot and Bahawalpur. There are other tribal groups who remain floating between the Ravi and the Satluj rivers. People living to the east of Multan, Pakpattan and Bahawalpur speak either Lehnda or Seraiki or both. The division between east and west Punjab is not due to language change. It is due to the concentration of the religious communities: Muslims, Sikhs and the Hindus. Both

Amritsar and Lahore play an important role of integrating the Punjabi identity based on the Punjabi language and culture independent of the political divisions. The Indo-Pak international border cannot divide the realm of the Punjabi and other dialects. Haryana was separated from the Punjab in 1964. People speak a Haryanvi form of Hindi in Kaithal and Kurukshetra, the latter being a pilgrimage centre of great importance to the Hindus.

From Lahore to Delhi—Guwahati

If one is travelling by road from Amritsar to Delhi one passes through well-known cities, such as Jalandhar, Ludhiana, Ambala and Panipat. The city of Delhi is located on the western bank of the river Yamuna. However, Shahadra, an extension of the Delhi territory, lies on the eastern bank. The Jats are numerically as well as in population a predominant culture group in this region. They speak a dialect, which is a mix of Punjabi, Urdu and the Haryanvi form of Hindi or Khari Boli. Yamuna, which flows on the borders of Haryana and Uttar Pradesh, is not a linguistic divide. As one travels from Ambala to Saharanpur or from Karnal to Muzaffarnagar there is no change in language. However, the Ganga, which forms the eastern boundary of the Ganga-Yamuna *doab* is a strong language divide. In this sector, Khari Boli is gradually replaced by Awadhi or by a mixed language, or a variant of it. A western variant of Awadhi is spoken in Bijnor, Rampur, Pilibhit, Bareilly, Badaun and Shahjahanpur. Further southwards, Kannauji emerges as the main dialect of communication. As one crosses the Ghaghara, also known as *Saryu* (Sarju) in the northeastern part of Uttar Pradesh, one notices a distinct change in language. Here Awadhi is replaced by Bhojpuri. The state boundary between Uttar Pradesh and Bihar is not a language divide. The districts of Siwan and Chhapra lie to the east of Barha. Here the Bhojpuri gives way to Maithili, a branch of eastern Hindi. But this may not be acceptable to the Maithilis because they claim Maithili as a separate language. In fact, they argue that Bengali, Assamese and Oriya languages have evolved from the Maithili. The Ganga has acted as a strong language divide between Maithili and Magadhi (also called Magahi) in Bihar. However, Purnia, Katihar and Bhagalpur form a kind of transitional zone between Bengali and Maithili. Districts, such as Araria, Purnia and Katihar lie close to the Indo-Bangladesh and Indo-Nepalese borders. There are significant Gorkha settlements along the Nepalese border with India. Here the spoken language changes from Nepalese to Maithili and Bengali. Darjiling, Jalpaiguri and Kooch Bihar form a zone of transition between Maithili and Bengali. Along the eastern border of Bihar a clear zone of transition is discernible between the Bengali and the Assamese speech forms. The concentration of certain tribal groups in this area has further added to the linguistic diversity. Gangtok, the capital city of Sikkim lies to the north of Darjiling. Due to its location between Nepal and Bhutan, it is clearly a transition zone and the people communicate in several dialects, particularly Gorkhali, Bhutia, Bengali and other tribal languages. East of the state of West Bengal, one reaches the domain of the Assamese language. The districts of Golpara, Kamrup,

Darrang, Guwahati and the neighbouring city of Dispur north of the Brahmaputra river constitute the core area of the Assamese language. However, the people of this area are bilingual, even multi-lingual. Bhutani language has its territory in Bhutan. This region lies to the north of the Assamese culture realm. The region south of Brahmaputra is characterised by multiplicity of dialects. The spoken languages include Assamese, and other tribal dialects. South of the Brahmaputra river the scene changes. The Meghalayan plateau with its uneven and rugged terrain has accommodated tribal groups who speak Garo, Khasi and Jaintia dialects alternately. They are multi-lingual. This helps them in day-to-day communication.

From Guwahati to Aizawl

The road between Guwahati and Aizawl passes through difficult terrain. Travelling between these two places, the road first takes one to Nagaon and then Dimapur (Assam) and Kohima (Nagaland). At Kohima the road bifurcates into two branches, the eastern one connects Kohima with Ukhrul and Imphal both in Manipur. The western branch passes through Senapati and Imphal. From Imphal the road goes to Churachandpur running in a westerly direction in the first stretch and then taking a turn to the south to reach Aizawl. If one takes this route to reach Mizoram, language changes swiftly from Assamese to Nagamese, Maithi, Manipuri, several Naga dialects and the Lushai. Travelling between Dimapur and Kohima, English is by and large the lingua franca. Besides English, variants of Naga dialects are the main spoken languages. From Imphal to Aizawl, Manipuri and Naga dialects slowly give way to the Lushai, the main dialect of the Mizoram. There is yet another road link which connects Guwahati with Shillong and Shillong with Silchar finally reaching Aizawl, the capital city of the state of Mizoram.

From Delhi to Mumbai

People travelling by rail between Delhi and Mumbai have several alternatives. One major route passes through parts of southern Haryana and northeastern Rajasthan. The broad-gauge line connects Delhi with Mumbai via Mathura, Agra, Gwalior, Jhansi, Bina, Bhopal, Itarsi, Khandwa, Bhusaval, Manmad and Kalyan. A second alternative available to the people is a meter-gauge line, which runs via Rewari, Alwar, Jaipur, Phulera, Ajmer, Chittaurgarh, Udaipur, Vadodara, Bharuch, Surat, and Silvassa to Mumbai.

According to the popular belief, the *bhasha/boli* changes at every 60 *kos* (original *kroh*). *Kos* is a measure of distance equal to approximately two British miles or 3.2 km. Thus one can say that generally the *boli* changes after every 300 km. In this distance of 900 km. the train passes through the following language domains: Bagri, Haryanvi (a form of Hindi-Urdu), Mewati, Meena, Mewari, Marwari, Bhili, Gujarati, Gujjari and Marathi. The Bhili has the following variants: Bhilala, including Bhil Garasia, Dholi-Bhili, Mewasi-Bhili, Rawal-Bhili, Tadvi-Bhili, Bhagalia, and Bhili, including Barela, Bhili-Minawati.

From Mumbai to Kochi

In this journey from Mumbai to Kochi the following language domains are traversed: Marathi, Konkani, Kannada, Coorgi, Tulu and Malayali. There are a number of tribal dialects spoken by Bhils and the Garasias. These groups are concentrated in their specific enclaves.

From Thimpu to Chittagong

Along this traverse, which represents the distance as the crow flies, several language domains are crossed, important among them being the Bhutia, Assamese, Bodo, Garo, Khasi, Jaintia, Bengali and the Chakma. Chakma is also spoken by the tribes of the same name in the adjoining parts of the Indian state of Tripura.

Ethnic Corridors

Case Studies of the Tribal Enclaves

It is commonly known that the Indian subcontinent is an example of unity in diversity. What is not generally known is the magnitude of diversity. It has been observed that the tribes in the mid-Indian region have a vast diversity in the spoken languages. An attempt has been made here to understand the ethnic diversity by selecting cross-sections at different points in space. We may recognise them as ethnic corridors. Keeping in view the fact that the mid-Indian peninsular region presents a vast diversity, it is necessary to make a journey across the region in order to explore the spatial patterns in the distribution of tribal languages and dialects.

Traverse 1: From Surat to Ranchi

This traverse passes through the districts of Surat (Gujarat), Dhule (Maharashtra), West Nimar, East Nimar, Betul, Chindwara, Seoni, Mandla, Shahdol, Sarguja (Madhya Pradesh), Ranchi (Jharkhand), etc. In all, there are 14 districts in this traverse.

Table 6.3: Tribal Concentration from Surat to Ranchi

Sl. No.	*Name of the District*	*Name of the Tribe*
1.	Surat	Bhil, Koli, Dhor, Kathodi, Naikda, Dhanka, Gamit
2.	Dhule	Bhil, Koli, Kathodi, Naikda, Dhanka, Gamit
3.	West Nimar	Bhil, Gond, Saharia
4.	East Nimar	Bhil, Gond, Saharia, Keer, Pardhi
5.	Betul	Bhil, Gond, Saharia, Kol, Pardhi, Pardhan, Nihal
6.	Chindwara	Bhil, Gond, Saharia, Kol, Bharia, Bhumij, Pardhi, Pardhan, Oraon, Parja, Korwa, Bhaina, Dhanwar

(*Contd.*)

(Contd.)

Sl. No.	Name of the District	Name of the Tribe
7.	Seoni	Bhil, Gond, Majhi, Kol, Bharia, Bhumia, Baiga, Pardhi, Pardhan, Kamar, Oraon, Halba, Halbi, Korba, Bhaina, Kharia, Bhunjia
8.	Mandla	Bhil, Gond, Kol, Bhariya, Bhumia, Baiga, Pardhi, Pardhan, Kamar, Halba, Halbi, Bhunjia, Parja
9.	Shahdol	Bhil, Gond, Nat, Sonr, Majhi, Kharir, Kol, Bharia, Bhumia, Mawasi, Baiga, Panika, Pao, Agaria
10.	Sarguja	Gond, Saharia, Majhi, Khairwar, Kol, Bharia, Bhumia, Baiga, Pardhi, Pardhan, Nihal, Kamar, Oraon, Halba, Halbi, Bhunjia, Parja, Dhanwar, Kharia, Munda
11.	Ranchi	Santal, Oraon, Munda, Bhumij, Mahli, Maheli, Ho, Kora, Malpaharia, Lohar, Lohra, Gond, Kharia, Kharwar, Bedia, Kamali, Korwa, Savar, Birhor, Gorait, Chik-Baraik, Khond, Bathudi, Kisan, Parhaiya, Binghia, Chero, Asur, Birjia
12.	Palamau	Santal, Oraon, Munda, Mahli/Maheli, Ho, Lohara/Lodh, Gond, Kharia, Kharwar, Bedia, Kamali, Korwa, Birhor, Gorait, Chik-Baraik, Bathudi, Kisan, Parhaiya, Chero, Asur, Birjia
13.	Hazaribagh	
14.	Santal Pargana	Santal, Oraon, Munda, Bhumij, Mahli/Maheli, Ho, Kora, Mal-Paharia, Lohara/Lodh, Kharia, Kharwar, Bedia, Karmali, Korwasavar, Gorait, Sauria-Paharia, Khond, Kisan, Parahaya, Birjia

Traverse 2: From Bhopal to Rangareddy

This traverse passes through Bhopal, Raisen, Hoshangabad and Betul (Madhya Pardesh), Amravati, Akola, Yavatmal, Nanded (Maharashtra), Adilabad, Medak and Rangareddy (Andhra Pradesh). This traverse consists of 11 districts.

Table 6.4: Tribal Concentration from Bhopal to Rangareddy

Sl. No.	Name of the District	Name of the Tribe
1.	Bhopal	
2.	Raisen	Bhil, Gond, Korku/Karku, Sahariya, Kol, Mawasi, Baiga, Panika, Pao, Agaria, Biar
3.	Hoshangabad	Bhil, Gond, Sahariya, Kol, Bharia/Bhumia, Biar, Keer, Pardhi
4.	Betul	
5.	Amravati	Bhil, Gond, Sahariya, Pardhan, Kamar, Halba/Halbi, Bhunjia
6.	Akola	
7.	Yavatmal	Bhil, Gond, Sahariya, Kol, Baiga, Pardhi, Andh, Pardhan, Nihal, Kamar, Oraon, Halba/Halbi, Bhumij, Parja, Korwa, Bhaina, Dhanwar, Bhunjia, Munda, Sawar, Kanwar, Kolam, Koya
8.	Nanded	Bhil, Gond, Andh, Pardhan, Kamar, Halba/Halbi, Bhaina, Dhanwar, Bhunjia, Kolam
9.	Adilabad	Bhil, Gond, Pardhan, Kolam, Koya
10.	Medak	
11.	Rangareddy	

Traverse 3: From Ratnagiri to Ganjam

In this traverse one passes through Ratnagiri, Sangli, Sholapur, Osmanabad (Maharashtra), Nizamabad, Karimnagar, Adilabad (Andhra Pradesh), Chandrapur (Maharashtra), Bastar (Chhattisgarh), Koraput (Orissa), Srikakulam (Andhra Pradesh), Ganjam (Orissa). There are 12 districts in all.

Table 6.5: Tribal Concentration from Ratnagiri to Ganjam

Sl. No.	Name of the District	Name of the Tribe
1.	Ratnagiri	Bhil, Koli Dhor
2.	Sangli	Bhil, Koli Dhor, Kathodi, Naikda
3.	Sholapur	Bhil, Saharia, Koli Dhor, Gamit
4.	Osmanabad	
5.	Nizamabad	Gond, Chenchu
6.	Karimnagar	Gond, Koya
7.	Adilabad	Bhil, Gond, Pardhan
8.	Chandrapur	Bhil, Gond, Khairwar, Pardhi, Pardhan, Kamar, Oraon, Halba or Halbi, Bhunjia, Karwa, Bhaina, Dharwar, Kharia, Munda, Sawar, Kamar, Kolam
9.	Bastar	Gond, Bharia/Bhumia, Pardhi, Pardhan, Kamar, Oraon, Halba or Halbi, Bhunjia, Munda, Saonta, Khond, Kondh, Bhattra
10.	Koraput	Gond, Koya, Manna Dhora, Gadaba, Parja, Bagata, Kondhs, Jatapus, Kotia, Benthoria, Munda, Shabar or Lodha, Bhuiya, Kolha, Kharia, Mirdas, Mahali, Matya, Pentia, Omantia, Banjari, Santal, Kora, Ho, Dharua, Holva, Lodha, Bhottoda, Sounti, Bhumia, Kisan, Jung, Parenga
11.	Srikakulam	Chenchu, Yerukulas, Konda Kapus, Manna Dhora, Savar, Hondareddis, Konda Dhora, Malis, Kulia, Kammara, Gondo, Gond, Nayaks, Mukha Dhora, Valmiki, Gadaba, Porja, Bagata, Kondh, Jatapus, Kotia Benthoriya
12.	Ganjam	Gond, Chenchu, Savar, Konda Dhora, Gadaba, Kondhs, Jatapus, Munda, Shabar or Lodha, Kharia, Matya, Santal, Bhunjia, Lodha, Souni, Bhumia, Didayi, Bondo

Traverse 4: From Koraput to Ranchi

The traverse takes one through Koraput, Kalahandi, Phulabani, Dhenkanal, Keonjhar, Mayurbhanj (Orissa), Mednipur, Bankura, Barddhaman, Birbhum (West Bengal), Ranchi (Jharkhand). The total number of districts covered is 11.

Table 6.6: Tribal Concentration from Koraput to Ranchi

Sl. No.	Name of the District	Name of the Tribe
1.	Koraput	Gond, Koya, Savar, Konda Dhora, Gadaba, Porja, Bagata, Kondhs, Jatapus, Kotia Benthoriya, Munda, Shabar or Lodha, Bhuiya, Kolha, Kharia, Mirdhas, Mahali, Matya, Pentia, Omantya, Banjari, Santal, Koru, Ho, Dharua, Halva, Lodha, Bhottoda, Sounti, Bhumia, Kisan, Jung, Parenga

(Contd.)

(Contd.)

Sl. No.	Name of the District	Name of the Tribe
2.	Kalahandi	Gond, Manna Dhora, Kulia, Kondhs, Munda, Shabar or Lodha, Bhuiya, Kolha, Mirdhas, Mahali, Banjari, Santal, Ho, Dal, Dharua, Binjhal, Bhottoda, Kol
3.	Phulabani	
4.	Dhenkanal	Chenchu, Savar, Kondhs, Munda, Shabar or Lodha, Bhuiya, Kolha, Kharia, Mirdhas, Mahali, Bhumij, Matya, Pentia, Santal, Oraon, Kora, Ho, Dharua, Binjhal, Sounti, Kisan, Bhatudi, Jung
5.	Keonjhar	Gond, Koya, Savar, Konda Dhora, Kondhs, Munda, Shabar or Lodha, Bhuiya, Kolha, Kharia, Mirdhas, Mahali, Bhumij, Korua, Matya, Pentia, Santal, Oraon, Kora, Ho, Dal, Kol, Sounti, Kisan, Mundari, Bathudi, Jung, Parenga
6.	Mayurbhanj	Gond, Koya, Savar, Porja, Bagata, Kondhs, Munda, Shabar or Lodha, Bhuiya, Kolha, Kharia, Mirdhas, Mahali, Bhumij, Korua, Matya, Pentia, Omantya, Santal, Oraon, Kora, Ho, Dharua, Binjhal, Binjhia, Lodha, Sounti, Kisan, Rajuar, Mundari, Bathudi
7.	Mednipur	Santal, Oraon, Munda, Bhumij, Mahli/Maheli, Ho, Kora, Mal Paharia, Mru, Lodha, Chakma, Rabha, Mech
8.	Bankura	Santal, Oraon, Munda, Bhumij, Mahli/Maheli, Ho, Kora, Lodha, Mech
9.	Barddhaman	Santal, Oraon, Munda, Bhumij, Mahli/Maheli, Ho, Kora, Mal Paharia, Mru, Lodha, Chakma, Rabha, Bhutia, Garo
10.	Birbhum	Santal, Oraon, Mahli/Maheli, Ho, Mal Paharia
11.	Ranchi	Santal, Oraon, Munda, Bhumij, Mahli/Maheli, Kora, Mal Paharia, Lohara or Lodh, Gond, Kharia, Kherwar, Kamali, Korwa, Gorait, Sauria Paharia, Kisan, Pahariya

Traverse 5: From Darjiling to Lohit

In this traverse one passes through Darjiling, Jalpaiguri, Koch Bihar (West Bengal), Goalpara, Kamrup, Nagaon, Karbianglong (Assam), Wekho, Mokokchung, Mori (Nagaland), Tirap, Lohit (Arunachal Pradesh). The total number of districts covered is 12.

Table 6.7: Tribal Concentration from Darjiling to Lohit

Sl. No.	Name of the District	Name of the Tribe
1.	Darjiling	Santal, Oraon, Munda, Bhumij, Mahli/Maheli, Kora, Mal-pahariya, Mru, Lorha, Chakma, Rabha, Mech, Bhutia, Nagesia, Garo, Lepcha
2.	Jalpaiguri	Santal, Oraon, Munda, Bhumij, Mahli/Maheli, Ho, Kora, Mal-pahariya, Mru, Lorha, Chakma, Rabha, Mech, Bhutia, Nagesia, Garo, Lepcha
3.	Koch Bihar	Santal, Oraon, Munda, Bhumij, Mahli/Maheli, Kora, Lorha, Chakma, Mech, Bhutia, Nagesia, Garo, Lepcha

(Contd.)

(Contd.)

Sl. No.	Name of the District	Name of the Tribe
4.	Goalpara	Boro, Boro-Kachari, Miri, Rabha, Mech, Deori, Hojai
5.	Kamrup	Boro, Boro-Kachari, Miri, Rabha, Mech, Deori, Hojai, Lalung
6.	Nagaon	Boro, Boro-Kachari, Miri, Rabha, Mech, Deori, Hojai, Lalung
7.	Karbianglong	
8.	Wekho	
9.	Mokokchung	Khasi-Jaintia, Naga, Garo, Mizo, Mikir, Demasa (Kachari), Kuki
10.	Mori	
11.	Tirap	Gallong, Tagin, Miniyong, Monpa, Mishmi, Padam, Khampti, Nocte, Wancho, Moglum Tangsa, Mossang, Tangsa
12.	Lohit	Gallong, Miniyong, Apatani, Hill-Miri, Mishmi, Padam, Khampti, Nocte, Pasi, Khamba

Tables given above indicate how the composition of the tribal population goes on changing. There are six tribal groups, which are inhabited in Dhule, seven tribal groups in Betul, eight in Chhindwara, Seoni, Mandla, Shahdol, Sarguja, Ranchi, Palamau and Hazaribagh. As the Table shows, while Bhils are the main tribal community, they are replaced by Gonds in Sarguja and by Santals in Ranchi, Palamau and Santal Parganas. From West Nimar to Shahdol, the Gonds are the major tribal community. They are replaced by Saharias in Sarguja, by Oraons in Ranchi, and Palamau. The number of tribal communities and their combination continues to change all along the traverse. It is evident that every district presents a highly diverse picture in terms of ethnic composition. Palamau by far presents the highest degree of diversity with the number of tribal communities being as many as 26. But this is not the end of the story. There are many other communities described as scheduled castes who also live in the same district. Each one of them, irrespective of the fact whether they are tribes or castes, have their own habitats in the district since they have carved out niches for themselves. There is no overlap. The enclaves in which they live have survived the onslaught of time. Perhaps time in itself is constant. One finds that the ethnic scene changes in space, not in time. It is important to understand that any planning at the level of the district will be meaningless without taking into account the ground reality. It is not the district but the community, which should be the target group.

The elements of socio-cultural diversity described in the above paragraphs are incomplete unless one takes note of the fact that there are scores of low castes or scheduled castes who also live in the same district but not together. Each caste group and the tribe have their own mutually exclusive enclaves. But there is no evidence to show that these groups interact with one another.

A Resume

The settlement of the primeval groups in Early India is an evidence of the ethnic structure of the people of the subcontinent. The present population of the subcontinent has been derived from a large number of ethno-lingual groups who came to the region at different points of time. The concentration of the racial stocks in the different regions of the subcontinent makes this point abundantly clear. The geographical arrangement of the racial groups is interesting. The population of the southern region displays strong affinity with the Palaeo-Mediterranean stock. The weft and web of the social fabric was woven by the Palaeo-Mediterranean stock, the main stream of the Mediterranean stock, and the so-called Oriental Group. On the other hand, the tribal population of the subcontinent shows elements of the Proto-Austroloid stock. The Indian tribes are distributed broadly in the Aravalli region, central Vindhyan complex—Sahyadriparvat, Vindhyachal, Satpura, and the Kaimur range. They are also concentrated in the Chotanagpur Plateau, in Chhattisgarh, Jharkhand, Orissa Highlands and Rajmahal Hills. The mountainous regions from Laddakh to Arunachal Pradesh and other regions in the northeast are the home of different strands of the Mongoloid stock. But these are generalisations. Perhaps the reality on the ground is far too complex.

The people of the northwestern region of the subcontinent have a strong affinity with the Nordic race. The northwestern region is also the core area of the so-called Oriental Group of the Mediterranean stock. The Pushtuns, the Kambohs and the other tribal groups of Afghanistan-Pakistan borderlands belong to the same stock. The people in the north and the northwest have affinity with the Palaeo and the Tibeto-Mongoloid stock in the sub-Himalayan region and the hills in the northeastern region have the same racial characteristics. The Proto-Austroloids are confined to the mid-Indian Vindhyan complex from the Aravalli and the Vindhayas to the Chotanagpur plateau and Chhattisgarh. The ethnic structure of population presents a mixed picture. It may, however, be noted that as we move from the periphery to the core regions of the subcontinent, ethnic distinctiveness loses much of its meaning since different racial types have merged with each other almost imperceptibly.

Ethnic characteristics of the population are the basis of regional differentiation. The regions have been recognised on the basis of ethnic criteria. It has been suggested that the Negritto stock, if any, was by and large assimilated into other racial groups. The remnants of this stock are found among the tribes of Andaman Islands as well as in the Nilgiri hills. Anthropologists suggest that the Proto-Austroloids were concentrated in the regions, by and large not suited for agriculture. The belt of tribal concentration extends from Surat in Gujarat and encompasses the Sahyadris, Aravallis, Vindhyas, Satpuras, Chotanagpur and Chhattisgarh Plateau up to the Rajmahal Hills. The quality of life and the level of human development of the tribal groups is very low. This should be a matter of concern. People belonging to the Palaeo- and the Tibeto-Mongoloid stock live in their small worlds within the mountains of the north. Similarly, people in the northeastern regions have lived in pockets of relative isolation for centuries. Their old technology has not enabled them to reduce their backwardness. In fact, the technology used by them is so primitive that it has arrested their progress. Their dependence on the sub-tropical forest eco-system has acted as a deterrent for their development. It does not offer opportunities for higher development. It may be noted that these lagging people are now being integrated in the democratic 'polity' of the nations of the subcontinent. The tribes of Bangladesh and Sri Lanka have similar characteristics. Pakistan is an entirely different case. The tribes there are not much different from the general population. It is noteworthy that the electoral politics has not acted as an effective catalyst of change. It has not been successful in the transmission of growth impulses leading to a comprehensive development. Despite official intervention the regional strains and stresses persist. They are a consequence of their economic backwardness. The gap between the tribal and the non-tribal elements of population is very large. It can be bridged only by minimising the disparities in economic development. The electoral politics notwithstanding, a systematic retailoring of the development programmes has not been done so far. The political economy has not succeeded in creating an environment conducive to development. The developmental benefits have not percolated to the deprived and the underprivileged sections of population. Both tribes and the general population have a similar experience. It may be stated that public expenditure on development projects is no guarantee of tribal development, just as mass participation in elections is no guarantee for democracy. The democracy reveals itself in the electoral rolls in the proportion of population not appearing in the voters list, and in counting the votes. When people exercise their right of franchise they have faith in the candidates to whom they vote. Their job is not over by casting the vote. In fact, their role as voters should continue even after the elections. There is no democracy if the elected candidates shift their loyalties from one party to the other. This implies that the elected candidates are not promoting democracy. When they do so they cease to be the people's representative. In this light one can argue that these events are not only anti-democratic, they lead to the disillusionment of the people.

END NOTES AND REFERENCES

1. The word society (*samaj*) is defined as 'a social way of living, customs, etc., of a civilised segment whereby people live together in organised communities'. The objective of organised communities does not apply to India, because the society is divided along caste lines, which is a discriminatory system based on social hierarchy (Srinivas 1962, 1986, 1992).
2. The concept of 'social order' implies organisation of societies who follow a civil code of conduct keeping in mind the context in which the population is organised in the subcontinent. The idea of social order is not transcendental to caste/community division.
3. The caste conclaves are attended by representatives of the same caste coming from hundreds of villages. They may be described as decision-making assemblies, where several choices are discussed. This procedure helps in the selection of candidates who are contesting elections on several political party tickets. '...in many parts of the

subcontinent lower castes were (treated as) serfs or slaves, either attached to the land and liable to be transferred along with it, or attached to the land owner or liable to be sold by him....But even now the agricultural hierarchy is mixed up in different ways and degrees with the caste hierarchy in several parts of India' (Srinivas 1962: 16).

4. The political party system in India since independence is more and more tilted towards the caste and community groups, since caste and community groups are distributed in geographical space. This fact is given weight not only in the selection of candidates but also in rallying support for them. The politicisation of ethnic, communal and caste identities has given a new lease of life to both division of votes among these groups and perpetration of caste prejudice.
5. The concept of commensality is applicable to the practice of giving and taking food from a certain caste or communal background and taking food by the members of the high caste or a community whether a Brahmin can accept food from a family who holds a high rank in caste. Similarly, giving food depends on whether the hierarchy is being disturbed or not.
6. The philosophy of secularism is based on the view that morality and education should not be based on religion. This means that any view of society will have to conform to the caste/community formation.
7. A racial group generally represented by aboriginal population of peninsular India. They are believed to be 'primitive'; they have a short stature and dark colour.
8. Archaeologists believe that the Baluchistan Neolithic culture was the oldest in the subcontinent. These Neolithic sites are located along the Indo-Iranian borderlands in Baluchistan.
9. Like Baluchistan, Sind is also a type area of the Neolithic cultural tradition.
10. There are passes all along the Himalayan mountain range. References have been made to the passes through which the people moved from central Asia, Tibet and the northeast to reach the subcontinent. The passes in the Hindu-Koh range in the Bamiyan province of Afghanistan are well known. The historically important Khybar Pass and the Bolan Pass in the Kirthar Range are examples of difficult passes. There are other passes such as Zoji La, Rohtang and Hamta.
11. By ethnic intermixing we mean a process of particular ethnic groups meeting each other. This leads to the emergence of hybrid varieties.
12. *Ba khal-e Hinduasht bakhsham Samarkand-o Bukhara ra*. The Persian verse quoted here reflects the popular feelings of an Iranian poet. It means 'I will bestow Samarkand and Bukhara, the famous cities of Uzbekistan, on the glowing cheeks of the Indian women'. The present generation of the Uzbeks may not like the idea of donating the centres of Uzbek civilisation to anybody other than the Uzbeks themselves.
13. It is suggested that the north Indian Plain was not a major attraction for the wandering people. The Plain was attractive for the ethnic groups already aware of the mode of agricultural production. This is an old hypothesis advanced by Subbarao. His concept of regions of attraction and regions of repulsion are well known.
14. Race is a biological concept. Every racial group is characterised by certain traits. If those traits which distinguish a certain group are found, we may say they are strands coming into being as a result of intermixing.
15. Reference may be made to the work of B.G. Tilak and other nationalist leaders of the right wing who postulated a theory that the Indo-Aryans were the natives of the subcontinent, that they were insiders and not outsiders. The issue is highly disputable.

16. The northern Himalayan and Hindu-Koh mountain complex is characterised by diverse ecologies and diverse cultures, the culture being a response to the specific ecologies.
17. B.S. Guha followed a scheme of classification of population into racial groups. Guha's approach appears to be a sweeping generalisation.
18. Herbert Risley was influenced by Nesfield and Ibbetson, who suggested that anthropometry may provide data for establishing the racial affinity of a group of the people (p. 115).
19. For Risley the morphometric data on caste groups may be considered to be important for establishing the ethnic identity (p. 119).
20. The idea of the ancient Aryans refers to the earliest streams of migration of the Indo-Aryans who are best represented in Punjab and other northwestern parts of the subcontinent (see Risley p. 119).
21. Indo-Aryans are supposed to be close to the Nordic race. Their physical characteristics are: powerfully built body, tall stature, long massive head, arched forehead and long face. The skin colour may vary from light brown to white hazel, grey and green. Their eyes are blue and the nose is long and narrow. There were other similarities, such as golden, blonde and blue hair. However, the tropical-subtropical conditions of northwestern parts of the subcontinent introduced changes in the original traits of the Nordic race. The reasons are multiple hereditary factors, genetic changes/mutations (p. 119).
22. 'Results mean that there is no correspondence between social gradation and physical differentiation in Bombay Presidency (see Ghurye, p. 120).
23. There is a relationship between ecology and economy. As indicated earlier the river valleys offered scope for agriculture, the surrounding hills and the plateau surfaces on the fringe of these valleys were inhabited by tribes (Ahmad 1999: 82, *Social Geography*).
24. Subbarao's concept of perennial nuclear regions, regions of relative isolation and regions of isolation is a neat schema. But it is too simplistic. The mechanics of interrelationship between the pastoral nomads and the peasants depended on agriculture which cannot be ignored. The primitive nomadism and the agricultural communities were affected by the people living in the river valleys. Most probably, the peasantry pressurised to squeeze the tribes out.
25. The displaced population from Afghanistan found refuge in the neighbouring countries of Iran and Pakistan. The fluid political situation lasted for decades and there was no alternative to out migration.
26. Achaemenidian Empire extended from Indus in the east to the Mediterranean seaboard in the west. A bulk of the territory of Turkey was also annexed by the Iranians. The region now known as Afghanistan was a part of the empire. Due to the location of Afghanistan it has remained exposed to the traffic of nomads and wandering people who were on the move from central Asia and west Asia to the subcontinent.
27. The Iranian Achaemenidian Empire annexed all parts of west of the Indus and other tributary streams of the Indus. The Achaemenidian Empire lasted from c. 559—330 B.C. (*Historical Atlas of Iran*, Tehran University, 1971, Plate 5).
28. The most important piece of evidence is the Persian language spoken by the Afghans and other tribes of modern Afghanistan.
29. By the time of Mauryan Empire under Asoka, the region came under the influence of Buddhism. In fact, the Kabul valley and the Bamiyan province of Afghanistan have inscribed the relics of Buddhism on the rock.

30. Seven hundred years later the entire population was Islamisised almost completely.
31. The United States of America and the United Kingdom joined hands to crush the Taliban government of Afghanistan. The idea was to relieve the people of Afghanistan from the clutches of Islamic fundamentalism.
32. The objective of allied bombing was to root out Taliban regime and capture Osama Bin Laden. Only the first objective could be achieved.
33. There is no authentic record of the size of the population of different ethnic groups. The estimated population as in 1979 is as follows: Pushtuns (7 million), Tajiks (3.5 million), Hazaras (1.5 million), Uzbeks (1.3 million), Aimaqs (0.8 million), Farsiwan and Hiratis (0.6 million), Turkmans (0.3 million), Brahui (0.1 million), Baluchi (0.1 million), and Nuristani (0.1 million).
34. The term Muslim should be replaced by Musalman. The name of the religion is Islam and its followers are known as Musalmans. The word Muslim appears to have been introduced by orientalogists of the western world.

7

Polity

In this chapter an attempt has been made to recapitulate the Hindu political tradition and its various versions. They have evolved during the 4,000 years of India's recorded history. Besides presenting an outline of the guiding principles of the Hindu polity, an attempt has been made to review the different approaches. One lesson, which a student of our political history should learn, is that we never lived in peace and harmony with our neighbouring kingdoms/states. Basham's hypothesis is proved by facts and it remains by and large close to truth.

India at independence on August 15, 1947 inherited a politico-administrative system, which was an amalgam of the political tradition handed over to us by the Greeks, Iranians, Greco-Bactrians, Mauryas, Guptas, Sakayas, Kushanas, Satvahnas, Rathors, Afghans, Khiljis, Tughlaks, Vijayanagar kings, Lodhis, Mughals and the British. The tradition was too complex to be narrated in a few words. Moreover, the tradition assimilated elements of the tribal groups, such as the Nagas, Khasis, Bhils, Gonds, Santals and the Oraons, to mention only a few of the tribal communities. The political tradition, which was received by the Indian state, had components of the Maratha, Sikh, Rajput and Telengana polity. The tradition was a synthesis of several regional traditions, old as well as new. The elements of Hindu, Jaina and Buddhist traditions merged with one another, often unrecognisably. The Indian state was influenced by the British who ruled over India when the country became a crown colony of England. The Constitution of India, as is generally known, was a modified version of the India Act of 1935, with of course, drastic changes in the make-up to give it a hue of democracy and socialism. The interim Government of India before attaining independence functioned within the limits of jurisdiction set by the British. It would be wrong to say that the Hindu polity was a unified tradition. In fact, there were several streams of the Hindu practices, which were flowing separately and never merged into a mainstream. The Mughals were not alone or outsiders all in all. Abul Fazal Allami and Raja Todarmal, Akbar's prime minister and finance minister respectively, based their politico-administrative set-up on the ground reality as perceived by them. The Hindu polity also derived the basic elements of the policies and practices of the kingdoms of South India, such as the Pandyas, Cholas, Bellals, Telenganas, Kalingas, the Vijayanagar kings,

Bahmini kings, Bijapur, Golconda, Ahmad Nagar, Bidar and Berar. There were several other models, such as the tradition of the Sikh rulers in the Punjab, Maharana Pratap of Udaipur, Shivaji and other kings.

The basic elements of the Indian polity were derived from the Hindu metaphysics, particularly the image of the Raja as an incarnation of the God Almighty. This image remained intact in the subconscious mind of the people of South Asia even after independence. However, Indian prime ministers from Nehru to Atal Bihari Vajpayee were elected democratically, but the post of the prime minister had elements of authority. The political party dictated its terms and did not allow the prime minister to function independently. The scions of the princely houses, like the Maharajas of Marwar, Amber, Mewar in Rajasthan and the Maharaja of Mysore in South India, are regarded as icons even today. The people of India considered the scions of the princely houses as reincarnation of divinity. It may be observed that the British government in the directly ruled provinces in the subcontinent followed as far as possible, the modern political tradition of Europe. The policies of the princely states, on the other hand, were essentially based on the principles of the ancient Indian polity. In a way, they were the carriers of the ancient Hindu political tradition. As is well known several Muslim princely houses adopted, by and large, the practices of the Hindu polity as modified by the British.

Origins of the Hindu Polity

History, we are told, is the art of record keeping—the record of happenings in the human society. The basic questions of the historian are simple: what happened, where and when? A more perceptive historian would add one more question: 'why' to his lexicon of what, where, and when. In answering the question 'why' historians have to subscribe to one or the other schools of thought: Marxian dialectical materialism, or idealistic Kantian philosophy of the critical period. Until recently, the historians generally were interested in the achievements of the ruling class—Kings, King Emperors, *Rajas*, *Maharajas* and the *Chakarvarti Rajas*.[1] They devoted their time and energies in collecting information about the lineage of the kings who ruled in different parts of the subcontinent. The genealogical tables, often kept by the *raj purohits*, served the basic purpose of maintaining a chronological order of the dynastic rule. That makes the record comparable with other similar records in other parts of the world or in other parts of the subcontinent. The subjects of these kings—the common people, men and women in the street, whose life was governed by these kings—lay, by and large, outside the framework of historical enquiry. Seen in the Indian context there were two basic elements on which the kings and the emperors were crucially dependent: first, a working class consisting of peasants and the artisans, and, secondly, the warriors. The warriors were absolutely necessary, not only to defend the territory from outside or inside contestants to power through encroachment, incursion or invasion, but also to make the production forces feel safe and secure. The ground reality was that the Hindu social order was the basis of the state. It was the responsibility of the *raja*, or the class of rulers, to protect the life and property of their subjects.[2] This was a responsibility of great significance assigned to the warriors, an organised armed force. Generally, the size of the army was broadly proportional to the size of the territory and its population. The king in his exercise of power supposedly followed a *Raj dharma,* or *dhamma,* to justify his position and strengthen his right to rule. The appropriation of a part of the surplus generated by the production forces, peasants and the artisans was the right of the king. The king legitimised his ruling status by linking it to divinity.[3] The act of ruling by the king was considered as an absolutely divine assignment, not questionable by the subjects. The warriors and the police, or home guards, were responsible for ensuring law and order. In adhering to the *Raj dharma*, or *dhamma*, the *raja*[4] or the king, depended on the functionaries of the state along with the chosen courtiers. The Mughal rulers, for example, had a number of *umraa* who justified his status as the ruler. The same *umraa* in the period before the Mughals behaved dubiously, by changing sides and supporting a certain individual of the Royal family to become the next king. If the *umraa* withdrew their support, the king had no alternative but to step down.

Indian history is full of such episodes when the *raja* assumed absolute power in order to strengthen his hold on the territory. But there were too many kings and too many kingdoms, and they were always at war with each other. Each dynasty had its own bards who kept the *raja* in good humour by making exaggerated statements in poetic form about the extent and boundaries of his kingdom and showering praise on the warriors, the so-called *surmas*. They made one and all to believe that in chivalry they were second to none. Each dynasty had its own *puranas* written and rewritten by each generation of kings. Indian history is also full of references to the king emperors, such as Ashoka (c. 269-232 B.C.) and Harsha Vardhana (c. 606-647 A.C.). These empire-builders rose from the petty kingdoms and the *janapadas* located at strategic points. They were in a better position to control the river routes and were responsible for the trans-shipment of goods and raw materials from one part of the country to the other. Their overlordship was evident from the fact that they granted the lower-order kings necessary freedom to continue with their local kingship or chieftainship, compromising on their sovereignty, while their authority as a suzerain power was acceptable to the local kings.

There were two basic elements in this process of empire-building in ancient India. First, the emperors subordinated the local kings, but they did not oust them from their nucleus of power. Secondly, they were forced to pay the tribute to the king emperor in proportion to the size of their kingdom. The tributary states were supposed to provide arms, men and horses in the case of an eventuality, or on demand. These empires expanded over vast areas in the Indo-Ganga plain.[5] Their expansion and contraction were cyclic in nature. The fertile alluvial plains drained by the Indus and the Ganga, favoured to a very large extent, the emergence and the sustainability of the empires. Their dismemberment was also quite common. The kingdoms and the people's republics (*Janapadas*) were located all over the country, in the sub-Himalayan region of the North as well as the plateau region of the South. Historians have noted that the rise of the empires, such as the

Achaemenid, Greek, Greco-Bactrian, Mauryan, Kushana, Satavahna, Khilji, Lodhi, Mughal, and the British was a phenomenon in itself. However, the empires were ephemeral and not permanent. On the other hand, the *janapadas* survived the onslaught of time irrespective of the fact that the king emperor[6] was authoritarian. One may say in conclusion that while the kings of the small *janapadas* were perennially in power, the king emperors were not. The empires were highly fragile. A careful study of the Bardic literature shows that the descriptions of the territorial extent and boundaries of the little kings were sometimes highly exaggerated and often mythical.

The Hindu polity was an extension of the social system. The *tika* or the *tilak* was applied on the forehead of the king by the *raj purohit*, not necessarily a Brahman, in an impressive coronation ceremony. Usually the *raja* was a Kshatriya. The courtiers of the *raja* were men of letters, poets, playwrights or scholars. The Vaishya and the descendents of the former *rajas* were, by and large, not included in this category. Very often, festivals such as the Holi and Dusshera were celebrated with great pomp and show. The king presided over these ceremonies and the distinguished courtiers were given a prominent position in the *darbar*. The courtiers and the ministers also functioned as legislators and had judiciary powers at the same time. The embassies of the rulers of the other countries of the world were also invited to the *darbar*. The *darbar* itself was a roving institution. It could be held anywhere in the kingdom. The sacrifice of the animals on these occasions was a common practice. The *raj purohit* was duty-bound to advise the king on the timing and place of the sacrificial rites. This practice continued in the families of the former kings. For example, the descendents of former rulers of Amber have a common practice of sacrificing a goat in the evening on every Thursday on the family deity.

Nation, Nationality and Nationalism

It is believed that the politico-administrative set-up was based on ethnicity and the tribal *jirga*/clan formation. The concept of clan or *jirga* is based on the extended kinship identity. But this concept was not regional or territorial. There is a difference between the clan and nationality. Nationality is a controversial concept. On the other hand, the concept of nation is more substantive. The problem of nationality became a debatable issue in the erstwhile Soviet Union. The Russian word for nationality is *narodnist*. Perhaps, there is no suitable word in Hindi and Urdu to describe the term nationality. The origin of this term, in our view, lies in the Latin words *nasco, natus, sum* and *nasci*, which means 'to be born'. The Latin word for birth is *natio-onis*. There is a close relationship between this word and the word *cognatio*, which means blood relationship; the Latin word is *cognatus*. There is a general agreement that the word *natio* is derived from the Latin *natio*, its French version is *nacion*. By implication, *natio* generally refers to human groups having blood relations and living in specific geographical space. It was known as *nacioun* in the medieval English, a synonym of *peple* and *pueple*. The word was derived from

Latin *populus*. In India, words like people and population are generally interchangeable. Similarly, the words racial and ethnic are interchangeable. However, the concept of race may be an expression of the purity or inter-mixing of groups.

The Greek presence in the northwestern part of the subcontinent for two hundred years appears to have left not much of an impact on the common people. However, there is some evidence to show that the ruling elites who represented the upper strata of the society were influenced by the *Hellenistic* tradition. Indian historian, Romila Thapar has mentioned about several communities who were living in specific geographical areas in ancient India. A particular reference has been made to Shakyas, Koliyas, Mallas, Yadavas and Vrijis. It is generally believed that Buddha (c. 563 B.C.—483 B.C.), who transformed the Hindu religious tradition, was born in a Shakya ruling family. On the other hand, Mahavira belonged to the Jñatrika'[7] tribe (Box 7.1).

Box 7.1: Invasion and Assimilation

The republican areas, by and large, remained less orthodox than the Ganges (Ganga) plain. Despite repeated attacks, isolated tribal republics continued to survive until the fourth century A.C. These were the areas in which Buddhism found considerable support, and where the foreign invaders—the Greeks, Shakas, Kushanas and Huns—were successfully assimilated (Thapar 1977: 53).

In some parts of the subcontinent, as records show, significant changes occurred in the territorial organisation. 'The battle for political pre-eminence continued'.[8] There were four rival states, three kingdoms (Kashi, Kosala, and Magadha) and one republic. On the other hand, Vrijis was a republic or a confederacy. 'The struggle for political pre-eminence lasted for no less than a hundred years' (Thapar 1977: 53).

Northwestern India was annexed to the Achaemenid Empire by c. 530 B.C. Emperor Cyrus crossed the Hindu-Koh range and established his authority with little or no resistance from the Kambojas and the Gandharas. At this time the capital of Gandhara was located at Takshashila/Taxila. The Iranians (Persians, Parthians) left an impact on the northwestern part of the subcontinent. Persian words and ideas were widely accepted. This is confirmed by several pieces of evidence: sigloi type coins; rock inscriptions of the Persian Emperor Darius (Dara); Kharoshthi script, widely used in northwestern India was derived from Aramaic. Alexander of Macedonia reduced the Iranian empire ending about 200 years of Iranian rule over the northern periphery of the subcontinent (c. 327 B.C.). While Alexander did not stay in the region, the Greek rule lasted for about 200 years. New political adjustments were made in the light of developments following the Greek invasion. While the masses were unaware of these developments, the ruling elites faced the consequences.

In so far as the subcontinent is concerned, democracy has been recognised as the basis of the state. But the experience shows that as a state policy on several occasions it has been abandoned and rendered dysfunctional. The future generations will determine their fate, fine-tune them or discard them as irrelevant. This may lead to their eventual death. Contemporary political trends indicate that it would be the market not federalism, which would unite the political space. The future generations of South Asians would create a new political ethos, discarding the ideals of democracy, federalism and secularism as irrelevant and out of tune with the times. The ideals of free trade would rule over the country and not the political agenda of the political parties. The trade would define and control the business territory. New centres of power are likely to emerge in consonance with the global trading systems. In recent years, globalisation has opened new vistas for trade agreements, treaties and contracts. Thinking globalisation would be a more logical strategy for change than thinking India. In our view the market place consists of three elements—*producers* of goods and services, *middlemen* who procure goods and sell them to the *buyers*. There will be no scope for any specific place emerging as the centre. The entire planet Earth would be the playground for these actors of the mercantile stage. Democracy can be successful in countries, such as Switzerland, United Kingdom and the United States of America. These countries have small populations and have been able to attain a high level of political development.

It is believed that the mosaic of Indian federalism is built on the foundation of language and dialect, religious communities, denominational sects, castes and sub-castes, regional and sub-regional structure, ethnicity and defined culture-patterns (Khan 1997: 9).

In the following paragraphs an attempt has been made to summarise briefly the major developments in the polity of the South Asian nations.

Afghanistan

The polity is an end product of the social structure and inter-linkages between different sections of population and economic sectors of production. The essential element in the Afghan polity is the *jirga* system. The inter-ethnic competition in controlling and appropriating the resources is the crux of the political problem. The people of Afghanistan faced a serious problem caused by the Western powers, particularly the United States of America and the United Kingdom, who jointly carried out operations to destroy the nation. The allied forces decided to do so without consulting the United Nations Security Council. October 7, 2001 will be remembered as a landmark in the history of Afghanistan as well as the history of the subcontinent. The occasion was the American massive retaliatory action against the Taliban regime and the US conviction that Osama Bin Laden was behind the attack on World Trade Towers (WTT) and that he was hiding in Afghanistan. The underlying assumption was that Afghanistan was a first step in the global war against terrorism. An important development was that Pakistan offered elbow space to the US forces. The retaliatory action came just twenty-six days after the attack on the WTT. The US-UK-led air strikes continued for roughly a hundred days. Taliban went underground; Osama and his Al-Quaida outfit could not be captured. The US did not apologise to the people of the world for massive military action taken on flimsy grounds. The war was fought in the name of peace. But actually it resulted in widespread devastation and loss of life. The war also resulted in the disintegration of the social order of Afghanistan. The US-UK air strikes on the country cannot be seen in isolation and may be described as the latest development in a long-drawn history of conflict spanning over two decades. Earlier, the two super powers—USSR and the USA—destroyed not only the pride of a strong Afghan nation but also destroyed whatever was left of the development infrastructure. The foreign interventions have brought Afghanistan to the bottom in the list of nations on human development ranking (Human Development Indices).[9]

The situation in Afghanistan should be seen in the context of the Soviet policies of expansionism and a bitter competition between the two super powers, which came in its wake. Because of its locational disadvantage, Afghanistan has proved to be the best venue for the rival powers to vie with each other. The Afghan problem should be seen in the context of regional interest groups, like Pakistan, India and China and in its international dimension. The future of Afghanistan will depend on how the international aid-giving agencies would react to the proposals to rebuild the Afghan nation. A concerted effort under the UN supervision would be needed for completing the gigantic task of national reconstruction. It is understood that in any plan for reconstruction, social, ethnic and religious factors would play a basic role.

It is unlikely that the situation would improve in the near future. Taliban, an Islamic formation, has been subdued now but not forever. They will come back. The problem has been allowed to further aggravate for the reasons too well known. The most unfortunate part of the story is that the conflict has been ethnicised. This is evident from the fact that the mass killings and torture of the innocent people of Afghanistan was perpetrated by the Northern Alliance, an inside military formation. In fact, the impact of the Taliban forces cannot be erased from the memory of the common people before the US intervention. The scale of the humanitarian tragedy that hit Afghanistan in the wake of military operations is difficult to describe. Forgotten by the West, there were already over one million internally displaced people prior to October 7, 2001. Less than three weeks later, the UN announced that seventy per cent of the population of the cities of Herat, Jalalabad and Qandahar had fled to neighbouring countries. Conditions in makeshift refugee camps on the Pakistan-Afghanistan border were, to say the least, appalling. Moreover, vast areas of the land were littered with mines left behind from the previous stages in the conflict. The actual number of Afghan men, women and children who lost their lives cannot be easily ascertained. The allied forces could not count the number of dead.

It is a puzzle that while Afghanistan's population consists of different ethnic groups, none of them have a majority in any geographical region of the country. However, Pushtuns constitute the largest ethnic group. The population composition shows that *Sunni* Muslims are in a majority. The *Shia* Muslims are a small minority. The non-Pushtun ethnic groups are conscious of their socio-cultural identity. Their language is *Dari* and not Pushto. Women have always been a disadvantaged group. It may also be necessary to understand that the twenty-two years of civil strife has resulted in the countless number of deaths. Intellectual and professional elements in Afghan society were the major target of the Islamic regime.[10]

There is no gainsaying the fact that the prolonged armed conflict has destroyed the economy. Afghans have a form of subsistence agriculture, which is their economic mainstay, with rain-fed wheat grown with barley as the main crop. Other crops include grapes, olives, and pistachios. They are grown in places where the weather conditions are favourable. Ecology being what it is, pastoral activities and livestock rearing appear to be the only means for survival. Agricultural produce, dry fruits and pistachios are the main export commodities. The country has natural gas reserves, which were exported to the erstwhile Soviet Union (now Russia). But this phase was too short. The on-going military conflict has forced the people to run away leaving behind their traditional agriculture high and dry.[11] Some reports indicate that smuggling has also been an important element in export trade (or Afghan Transit Trade—ATT).[12]

The Soviet occupation of Afghanistan resulted in serious consequences. First, it led to mass exodus of the Afghans—3 million fled to Pakistan and almost the same number to Iran. Those who migrated to Iran were supposedly well-versed in the Persian language. Secondly, a vast chunk of the local population shifted to caves located in the mountains all around. They had minimal access to food and shelter. Pakistan's effort to establish a Mujahidin government in Kabul did not succeed initially. The socialist government of Afghanistan fell in April 1992. In fact, Soviet troops had left Afghanistan three years earlier. It is a paradox that the Soviet troops returned to their country where the socialist regime had become a thing of the past. The leader of this revolt was a Tajik commander, Ahmad Shah Masood, who operated from Kabul. As one of the solutions, Pushtun Mujahidin parties agreed to form a coalition government. But the infighting continued for three years or more, reducing much of Kabul to rubble.[13] Evidently, Hikmatyar, leader of the Mujahidin movement, failed in his efforts of establishing a functionally viable Mujahidin government. The Taliban took the benefit of this move.[14] It was reported that since 1994-95 the Taliban were holding sway over the southern parts of Afghanistan. Subsequently, they moved to non-Pushtun areas, which led to a new inter-tribal conflict.

There were situations in which the different factions of the Mujahidin parties played real politic. They failed to remember that economics was larger than politics. Their pre-occupation with *jihad* and the related issues did not allow them any chance to think of economic development.

Afghanistan has passed through several phases of outside intervention. The present borders of Afghanistan are only a century-old. In the ancient period it was annexed to the Iranian Empire and later ceded to Hellenistic Empire. The present day Afghanistan has evolved as a state in the middle of the eighteenth century when Ahmad Shah Durrani, a prominent Pushtun leader, conquered the territory and established its stronghold on the Afghan soil. The legitimacy of this arrangement was approved by Tsarist Russia and the British Empire in India in 1891, 1893, 1895, and 1896. In 1929, king Aman-Ullah was overthrown by a Tajik leader, Habib-Ullah II. Afghans were seriously concerned for their territorial integrity in the fifties of the last century. The two factors on which Afghans were seriously concerned were: (a) Soviet occupation and (b) Islamic regime, which resulted in far-reaching consequences. The year 1978 was a landmark in Afghan history when Peoples Democratic Party of Afghanistan (PDPA) succeeded in having control over the entire territory. This happened with full support from the Soviet Union. Both the British and the Russians were concerned on the rise of these rival political forces. In their wisdom a hostile regime should not come to power. The neighbouring country of Pakistan felt threatened by the Soviet expansion.

Pakistan

Pakistan became a sovereign country after the British Empire was partitioned in 1947. The justification was that the Muslims were concentrated in the western and northwestern parts of undivided India. The claim that these regions were Muslim-majority areas was based on the geographical distribution of the 'demographic' Muslims. The basis was the 1941 census of India. The two dominions of India and Pakistan were born in the mid-night of 14 and 15 August 1947. Pakistan consisted of two wings—East and West. East Pakistan was born when a movement started for separation of the Muslim Bengali majority areas. Partition was not a simple matter. Families were divided, homes and hearths were divided, roads and railways were divided, banks, assets and liabilities were divided. Partition brought devastation on a large scale unprecedented in the history of the subcontinent. Nationalist leaders and freedom fighters, such as Khan Abdul Ghaffar Khan and his associates in the Indian National Congress Party were disillusioned because they were living in the newly created state of Pakistan. Millions of people suffered. The nationalist leadership in Punjab, Sindh, Baluchistan and the North Western Frontier Province suffered a major setback. The division of the cultural heritage was even more shocking than the division of the land. Those who migrated to Pakistan left behind the memories of Delhi's Jama Masjid, Agra's Taj Mahal, Imam Baras of Lucknow, Fatehpur Sikri, red forts of Agra and Delhi, and the *dargahs* of Chishti saints, Hazrat Muinuddin Chishti of Ajmer, Hazrat Bakhtiar Kaki, Hazrat Nizamuddin Aulia and Amir Khusro of Delhi. Those who migrated from Pakistan to India left behind Nankana Sahib, the birthplace of Guru Nanak. Ajanta, Ellora and Taj Mahal were in India while Harappa and Mohen-jo-daro were in Pakistan. Hundreds of thousands of men, women and children were lost on the way to the new dominions of India and Pakistan. The violence that hit the people during the campaign for Pakistan and the partition caused colossal loss of human life. Murders and rapes were rampant. Women were taken away by force. Their parents, children and spouses were not traceable. They are not traceable even today. For the mothers, the memory of the lost children was a lifetime agony. The heritage of Tagore was on this side of the East Pakistan; Nazar-ul-Islam was on the other side of the newly demarcated international border between East and West Bengal. After the emergence of Bangladesh, the situation improved a little with both the nations recognising Tagore and Nazar-ul-Islam as part and parcel of a common Bengali heritage.

Did the partition of the British Empire in India solve any problems? We know that it is an open question. It was argued that the division of the subcontinent was necessary for the simple reason that a section of the Muslim population suffered a psychological shock mainly due to the imagined threat of Muslim subordination by the majority community. Whether Pakistan was a necessity for the fulfilment of the imagined aspirations of the Muslim community of the subcontinent is a disputable issue.

The newly-created state of Pakistan is now a reality and its status as a sovereign state is no longer questionable. Today, these issues are meaningless. Both India and Pakistan recognised their role in retaining a common identity of the people of the subcontinent. Today, the position is that Pakistan is a sovereign state, so is India and the people of the two countries look forward towards a future of peace and development in all the three nations of the subcontinent: India, Pakistan and Bangladesh. A sizeable chunk of the population who were demanding Pakistan on the basis of 'demographic majority' of the Muslim community, are now living in camps in Bangladesh, leading a miserable life. They do not have a future. Others belonging to the same category of refugees are living in Karachi and other parts of Sindh Province of Pakistan. They have to compete with better-placed sections of the population.

The country has the potential to enhance the volume of exports, but the political situation has often proved to be a major setback. As we already know Pakistan has passed through a series of *coup d'états*.[15] It has been argued that the military dictatorship has helped the people more than the democratically elected governments. But the most unfortunate part of the story is that democracy-dictatorship syndrome is not going to change. Democrats like Zulfiqar Ali Bhutto, or his daughter, Benazir Bhutto of the Pakistan Peoples Party, or Nawaz Sharif of the Pakistan Muslim League, who served as prime ministers at different stages in the history of Pakistan, did not succeed in their efforts to lay the foundations of a real democracy, fair, developing, and socially just. For example, *Muhajirs* (refugees from India) found military dictators safer than the democratically elected politicians. Governance of a country like Pakistan demanded an uncompromising commitment to the goals of the state as set up by the democratic process. It is the responsibility of the administrators to ensure law and order facilitating overall development, reconstruction of economy and implementation of the socially sensitive policies, particularly in the field of human development, education and health. The reality is that the pro-democracy leadership has remained busy in preserving their own interests more than the national interests. The dictatorship was supposedly more helpful to the weaker sections of the country. This cannot be said about the pro-democracy political leadership.

Bangladesh

The new Bangladeshi nation had to face the consequences of its extraordinary birth, its burgeoning population, economic underdevelopment and the crisis of identity being the major issues. While the UN agencies working in South Asia, helped the nation overcome the shocks of the caesarean birth of the nation, the task was too difficult, the people were too many and the economic growth rates too short of expectation. The young nation could not wriggle out from the poverty-underdevelopment-death syndrome. Peace remained elusive. It was but natural that the seeds of democracy could not sprout in the deltaic mud. The history of Bangladesh is the history of *coup d'états*. Sheikh Mujib-ur-Rahman, leader of the Awami League, was assassinated within a few months of his assuming the presidential office. The political instability proved to be the largest menace—twelve governments in twenty years.

It is noteworthy that a population of 150 million is crucially dependent on the 9.4 million hectares of the arable land. However, in actual effect, only 8.97 million hectares of land is brought under the plough in any normal year. But the normal years are not too many. The average density of population—1045 persons per sq. km.—is a crude index of population pressure on land. Much higher densities have been reported from the rice and jute producing areas along the banks of the Padma and the Meghna rivers. Things, as they are today, do not suggest further scope for any increase in the arable land. Demographic data on Bangladesh is indeed disturbing. Besides high densities of rural population, economic development has been lagging far behind the population growth. Bangladesh is a good example of men's struggle for survival. Perhaps the Malthusian arguments are more valid here than anywhere else in the world, barring, of course, some African states.

Nepal

The Chinese pilgrim Huen Tsaung visited Nepal in early seventh century A.C. It is reported that Huen Tsaung proceeded to the northeast and reached Vriji, which is 83 miles (approximately 133 km.) away from Vaishali. The kingdom has been identified as Wajji or Vriji. In Buddha's time, Vrijis were divided into several clans, like Lichhavis, Vaidehis and the Tirabhuktis. The Chinese pilgrim estimated the distance from Vriji to Nepal as around 250 miles (approximately 402 km.). There is some confusion about the actual location of the Nepal kingdom. Sometimes it has been identified as located in the Sapta Kausiki region, meaning seven streams of the Kosi river. According to Cunningham, Nepal occupied the basins of Kosi and Gandak. Early history of Nepal is, however, shrouded. The existing evidence shows that the *Raja* of Nepal, Ansu-Varmma, belonged to the Lichhavi clan. In local histories he is mentioned as Anghu Varmma. According to the tradition he belonged to the Newarit or Newar dynasty. The Chinese pilgrim visited Nepal in 637 A.C. It is understood that he was in Nepal towards the end of Varmma's reign. Perhaps the Nepalese history begins in 4 B.C. The first Lichhavi king conquered the territory of Nepal in 50 A.C. It is believed that the Lichhavi conquest of Nepal must have been preceded by 37 reigns before Anghu Varmma. The first king was

Newarit. Historians are not very clear about the later history of Nepal. We are told that the Rajput military caste from north India overran Nepal in the early fourteenth century. His descendents ruled over Nepal up to 1768 A.C. when the invading Gorkhas seized Nepal. They consolidated their power and attacked Tibet. However, Chinese army defeated the Gorkhas and occupied a part of Nepal for a short period. In 1803 a treaty was signed between the Gorkhas and the British. The situation suddenly changed when the British declared war on Nepal and defeated the Nepalese forces. A peace agreement was signed in 1815. According to this agreement the British demanded extensive parts of the *Terai* and the border areas, which were handed over to the British. Around the same time Nepal came under the political domination of the Rana family. Sir Jang Bahadur who himself was pro-British became the king. A peace treaty with Tibet was signed in 1856. According to this treaty the Rana king of Nepal asked Tibet to pay yearly tribute in lieu of peace. In 1857 when India was struggling for independence, it is understood, that the rulers of Nepal sided with the British. The British granted independence to Nepal on November 7, 1950. In another development, Maharaja Mohan Shamsher Rana removed the King Vir Bikram Tribhuvan who fled to India. Thus the contemporary political history of Nepal begins with 1950.

Sri Lanka

A legendary Prince, Vijaya conquered the island of Sri Lanka in 504 B.C. and subjugated a local aboriginal group known as the Veddas. It is believed that Vijaya was a Hindu king who came from northeastern region of the Indian subcontinent. The prince named the kingdom as Sinhala. He introduced commercial crops and encouraged Tamils to migrate to Sinhala for work on the plantation crops. Subsequently, in the third century B.C., Sinhala population adopted Buddhism and declared it as the national religion of his kingdom.

Sinhala was under the control of the Tamil kings for a long period of time beginning with the 3rd century A.C. to the middle of the 12th century A.C. The Chinese forces occupied the island for a brief period of thirty years. After the Chinese occupation ended, the island was divided into petty kingdoms. This happened in about middle of the 15th century (1439 A.C.). It was the period when the Portuguese were operating off the west coast of India and Ceylon (now Sri Lanka). They were looking for some space to anchor their ships to strengthen their trade interests in the region. The Portuguese built a fort at Colombo and developed Colombo as a trading port for their maritime trade in southern and southeastern parts of Asia. However, the competition between the Portuguese and the Dutch took a new turn when the latter threatened the Portuguese. By 1658 the Dutch were in full control of the island, the only exception being the kingdom of Kandy. Towards the end of the 18th century, the British started taking a keen interest in the affairs of the island. In a brief period of four years from 1798-1802, the British occupied the entire island including the Kandy kingdom. The Sinhala was formally ceded to Great Britain in 1803. The British named the island as Ceylon, which appears to be a corrupt version of the word

Sinhala. In 1815 Britain declared Ceylon as a crown colony. The struggle for independence started in early 19th century. The Sinhalese waged a war against the British. Several well-organised rebellions marked the Sinhalese dissent and opposition to the British presence in the island. They were not successful but the struggle for freedom further intensified. During the period 1817, 1843 and 1848, the island passed through a phase of civil strife. In 1931, a new constitution was promulgated by the British, the highlight was that the new constitution was granted a semi-autonomous status to Ceylon. However, autonomy was given in the internal affairs only. During the Second World War the island became the main base in the military operations against the Japanese. The island won freedom in early February 1948, with D.S. Senanayake as the first prime minister of the nation.

India

India after attaining independence in August 1947 received the heritage of a dismembered state. There were three main issues before the new nation and the government. The first priority was to integrate the different segments of political space into a single unified country. For achieving this goal the country launched a comprehensive programme of national integration. The seven hundred odd princely states were asked to join the Indian Union. This programme was implemented by Sardar Vallabh Bhai Patel, the then Home Minister of India. With his faculty of convincing and persuasion, Sardar Patel achieved success in the monumental task of building India as a nation-state. The second item on the agenda was to remove the disparities in regional, economic and political development. The country had faced long spells of drought leading to famines and death on a large scale. Even food grains had to be imported from outside. Thirdly, evolving a harmonious programme for the congenial relations between the Centre and the States was considered as the most important requirement for the country. Nehru emphasised the need for national integration. The Constitution of India was adopted in 1950. The Constitution recommended a form of federalism as the basis of the relationship between the Centre and the State.

The South Asian nations have faced this question time and again whether they are working for nation-building or federal nation-building? This question can be answered only after a careful study of the text of the Constitution of India. However, the constitution is specific to India and not to the subcontinent as a whole. A union of regional states/countries of the subcontinent can perhaps be the highest form of federalism. The day is not far away when the South Asian nations would constitute themselves into a confederacy of states.

This is besides the point that one can have either democracy or a secular state ensuring social justice. Social justice requires a monitoring authority strong enough and capable of imposing normative conditions for achieving this goal. Paradoxically, if one imposes anything from above (read Central/Union government) it is interpreted as authoritarianism and not democracy (Box 7.2).

Box 7.2: Federalism in India

The Constitution of India leaves gaps in information. Political scientists often comment that the document is on federation and not on federal nation-building. There are three determinants of federalism: first, federalism as a social-cultural theory. This recognises pluralism as a valid basis of ... peaceful coexistence, secondly, federalism as a political principle, which seeks to stabilise ... on the pattern of constitutional diffusion of power, and thirdly, federalism as an administrative arrangement to coordinate the legitimate distribution of power between the Centre and the State. In its praxis federalism builds and sustains the polity, which may be called *unity of polity* and simultaneously preserves and promotes the plurality of the society. In that light it is a composite nationalism, particularly working for democracy, secularism and social justice (Khan 1997: v; 1).

Indian state today is described as a federal nation. The general view is that the country with twenty-eight states and seven Union territories with variegated social structure, multiplicity of ethnic and ethno-lingual groups, a fast growing population and vast disparities in development, can function as a federal nation only. This raises the question: is India a confederacy of states? If yes, what are the principles or guidelines on which the system is functioning? While federalism calls for a balanced distribution of power between the union and the states, the political history since independence shows that many a time the centre has treated the state as taken for granted. Evidently, it has not been all smooth sailing. Political scientists consider federalism as the essential principle for organising heterogeneous societies into a viable pattern of sharing of political power by reconciling the twin processes of political unification and social diversity (Khan 1997: 3). It may be recapitulated that the issue of federalism was widely debated. This question has been asked during the last sixty years since independence and even before independence. It has been observed that there were two forces contradictory to each other—centripetal force and the centrifugal force. It is a physical law that on the periphery of the country centrifugal force is in excess of the centripetal force. The experience is not unique to India. The union government has always tried to balance the centrifugal force by generating a compatible centripetal force. In extraordinary situations the union government has acquired more power than sanctioned in the federal constitution.

'In a federation the equation centralism and decentralism between the union and the states is not dichotomous. It is a case of convergence, a pattern of cooperative distribution of jurisdiction and power.' Evidently, if there are no states/regions there is no federation. The federal principle is a new experience for India. As pointed out earlier ancient India was a highly politically fragmented part of the world. There were hundreds of patriarchies, monarchies, local/rural republican formations, feudal states and tiny chieftainships. In the process of evolution of the states the river valleys of the subcontinent emerged as the foci of settlement and of political power. It gave birth to *sultanates* where feudal elements ruled. Particularly this is true for princely states (*rajwadas*). In the history of the subcontinent there were a few phases of empire-building too (this issue has been discussed in chapter 5).

Political historians recollect that the whole of India was never unified into a single state ruled from a single centre of power, be it Patliputra, Kannauj, Indraprastha or Taxila. However, there were three exceptions to this general rule. The three or four examples may be cited here—the Mauryan Empire of Chandra Gupta Maurya and Ashoka (269—232 B.C.); the second empire under Sultan Alauddin Khilji (died 1316 A.C.) and the third under the Mughal kings from Babar to Aurangzeb (1526—1707 A.C.); and the British Empire. However, Mughals continued to rule after the death of Aurangzeb. But none of his descendants were capable of controlling such a vast territory. There is a Persian saying: *mumlakat-e-Shah Alam az Dilli ta Palam*, which means that one of Aurangzeb's descendants, Shah Alam, controlled a very small part of the country, perhaps from Delhi to Palam now a suburban neighbourhood of New Delhi. His jurisdiction was reduced by the other rising powers of the eighteenth century. The Mauryan Empire lasted for a brief period (321—185 B.C.). The decline of the Mauryan Empire is described as the end of the Hindu imperialism in the subcontinent. The Mughal power reached its zenith under Aurangzeb whose empire encompassed most of the area south of the Vindhyas. It also included parts of Afghanistan and the southern regions of central Asia east of the Kopet Dagh mountains and north of the Oxus river.

On looking back it appears that the British imperialism played the most constructive role in promoting the all-India consciousness, even though the subcontinent was being governed from outside. It is also important to realise that the nationalist movement demanding full independence was a direct response to the extent and boundaries of the British Empire. The wider the area under the British, the more people joined the brigade of the freedom fighters. It was the *Raj*, which promoted the consciousness of an all-India identity. In fact, the British helped the Indians indirectly to become conscious of this larger entity as compared to other criteria, such as caste, religion, ethnicity and language. It was this consciousness, which transformed the Indian value system. Regionalism was transformed into nationalism, parts became conscious of the whole, linguistic nationalities became conscious of their higher-level identity of nationalism. It was in this environment that the ideals of democracy, socialism, egalitarianism and federalism were nurtured. It was not easy for nationalist leaders to convert the self-centred feudal lords into a democracy. Democracy meant giving equal rights to all irrespective of their caste, language, religion and social status/rank, region or geographical location. In a theoretical sense democracy could steam-roll the hierarchical social structure. In practical terms it could not achieve the goal!

The polity of the south Indian kings was characterised by certain features. First, the king did not expect much from his state. It was his responsibility to protect and strengthen the existing social order and defend the interests of the people. In lieu of his services he received one-sixth part of the land produce. The social order was based on *sruti*, *smriti* and *acara*. The ruler did not interfere with the religious practices of the people and social and economic practices. An important feature of the kingdom was not to intervene in the day-to-day life of the people. But justice was to be ensured. There were other organisations, such as local leadership, caste groups, occupation and/or religious persuasions, which remained intact. There were autonomous groups and associations who took care of these problems. These groups functioned within the framework of age-old customs and practices. There used to be a general assembly and historical records refer to the village assemblies. The king had no control on these institutions. Secondly, the Kshatriyas were appointed to protect the people. The court consisted of individuals drawn from men of letters and the arts. The people in common perceived that aggrandizement was the duty of the king. The image of the king was that of a *Vijigishu* or a conqueror. The king followed a code of conduct and he was expected to maintain *status quo*. The structure of the polity was such that no radical political change could take place in the traditional polity. It is because of this feature that sudden or revolutionary change was not possible. On the other hand, in other regions of the world violent changes occurred. References have been made generally to the full-fledged autonomy and the self-sufficient village community. But these conclusions may not be viable in all situations. In the rural areas of the kingdom day-to-day life of the people was not affected by the ruling class or the *raja*. Two institutions were critical: (a) *muttuda*, sometimes described as *kilan* or *gramabhojaka*; and (b) village *sabha* or assembly. Urban places had traditional guidelines, for example, *nagaram* or the city council had special functions to perform where traders and merchants controlled the social life. However, the social life was determined by the traditional code of conduct.

These remarks apply to the general situation but nothing could stop the *raja* from leading a lavish life and celebrating his authority. Wining and womanising were common features. Historical records indicate that the Bahmini kings of the Deccan also indulged in these practices and happily maintained a *harem* consisting of wives and concubines. It will not be correct to say that these practices were permissible within Islam. Kings are kings and not Hindus or Muslims.

The South Asian Drama[16]

We have discussed in the following paragraphs experience of the South-Asian nations in the recent political history. The developments indicate the direction in which the polity was moving. What happened in the recent past is nothing short of a drama. Special reference has been made to Afghanistan, Pakistan, Bangladesh, Nepal, Sri Lanka and India.

Afghanistan

The Afghan viewed their nation as a monarchy and not a nation-state. It was the monarchy with which the people identified themselves. At the lower level their identity was expressed in the *jirga* or clan formation. The *jirga* was more specific than the monarchy. In the Afghan consciousness the monarchy was a summation of the different clans and sub-clans. Eventually, the king became a dynastic head and the political space controlled by him was his kingdom. His status is that of a despot and he carries with him the image of a saviour, not a contestant to power. His kingship is the domain of rule and carries with him a divine disposition. The Kingdom of Afghanistan was a symbol of consensus. The Russian occupation of Afghanistan was an aberration and a serious departure from the Afghan view of the state. It shattered the basis of the Afghan authority. The people of Afghanistan did not succumb to the pressure of a foreign regime. The Russian domination was described and perceived as a threat to Afghan security. They failed to identify themselves with the foreign regime. The main problem was that the Afghan resistance to this development resulted in an endless internecine war. The Russian dominance was not acceptable to the people. The struggle between the rival clan leaders aggravated further, when the US agencies incited the *jirga* leaders to oppose Soviet occupation of their territory. Evidently, the US agencies took advantage of the situation. This development disturbed the social order, particularly the inter-*jirga* relations.

The US converted Afghanistan into a cold-war theatre. The US armed the anti-Russian guerrilla groups to wage a war against the Russians and described the struggle as a *jihad*. The US agencies pumped in arms to pull down the Soviet regime. However, these arms were used not only against the Soviet army, but also disturbed the inter-clan balance.

The history of modern Afghanistan started with the establishment of the Durrani Empire in 1747. The country remained under the control of the British up to 1919. Reference has already been made to Anglo-Afghan wars. The country gained freedom from the British rule in 1919. However, the assassination of Nadir Shah in 1933 paved the way for the ascendancy of his son, Zahir Shah who became the next king. The king of Afghanistan, Zahir Shah, remained the head of the state for about four decades from 1933-1973. A newly established Peoples Democratic Party of Afghanistan (PDPA) was established by Babrak Karmal, Hafizullah Amin and Noor Mohammad Taraki in 1965. Daoud Khan, Zahir Shah's brother-in-law, who had a status equivalent to the prime minister under Zahir Shah's rule, overthrew the king and assumed the office of the president as well as prime minister in 1973. In 1976 the fourth Constitution was adopted, making Afghanistan a socialist state. But in 1978, Daoud Khan was assassinated and PDPA leader Taraki took over power, making Afghanistan a democracy. Daoud Khan introduced several progressive reforms including land reforms. The reforms, however, were based on the Marxian philosophy in right earnest, but in the course of implementation of his programme he became unpopular. The opposition forces united to throw him out. In September 1979 Hafizullah Amin ousted Taraki and in December of the same year, the Soviet government intervened and ousted Amin. Karmal was brought to power the same year. In March 1979 anti-government Muslim fundamentalist guerrillas led the country to civil strife. In 1986, Najibullah, a former head of the intelligence agency assumed the office of the president. A year later Soviet Prime Minister, Shevernadze, visited Kabul in January 1988 to discuss the issue of the future of the country. He put forward the official proposal for the peace treaty. The treaty was signed in Geneva in April 1988. The signatories included USA, UN, and the Soviets. In February 1989 Soviets withdrew their forces. But this created a kind of political vacuum leading to an internal war between government forces and anti-government guerrillas.

In April 1992, the communist regime was ousted from power. In 1996 the government was overthrown by Taliban, a fundamentalist Islamic organisation. The bloodshed continued as several ethnic groups were on the confrontation path. The situation became worse when the US decided to attack on Taliban centres through a powerful air attack. It took a very long time to stabilise the state. The explanation was that the September 11 attack on the World Trade Towers in New York was a handy work of Osama Bin Laden and his Al-Qaida outfit. The post-war Afghanistan is still to return to its normal state. The US asked the Afghan people to elect their own president. However, before the elections were held in 2004, Hamid Karzai was appointed as the president of the interim government.

Table 7.1: Afghanistan: Evolution of Polity

Time Period	*Political Events*
1973	Monarchy overthrown; Proclaimed as a Republic.
1978	Constitution suspended through a *coup d' état*; A revolutionary council formed with Soviet support.
1987	New Constitution adopted by People's Democratic Party.
April 1992	Communist regime ousted.
June 1993	Interim government takes over under a prime minister.
September 1996	Government ousted by Taliban, a fundamentalist organisation.
September 2001	US-UK launched a powerful air attack on Taliban centres; Occupy Afghanistan under a new regime.

Pakistan

The new nation-state of Pakistan was born on August 14, 1947. But it faced the consequences of a bloody struggle, which led to the partition of the British Empire in India. Millions of people were displaced. Trains leaving for India or leaving for Pakistan from India were attacked; women and children were kidnapped. The politics of hatred did not leave any scope for reconciliation. The violence spread like a hurricane. Neighbours attacked neighbours because they belonged to two different communities. The crazy mobs inspired by communalist politics had a free hand to play with. It is estimated that 3.5 million Sikhs and Hindus migrated to India. An estimated 5 million Muslims living in different towns, cities and villages of India migrated to Pakistan. Pakistan became a reality, but on the ashes of long-cherished ideals of Hindu-Muslim unity and solidarity. There were other related problems, which came along with the partition. In the first place, Junagarh, a princely state with Hindu majority located in the Kathiawar peninsula of Gujarat declared merger with Pakistan. This was not acceptable to the Indian government. In fact, Indian forces annexed the territory through an armed intervention. The Nawab of Junagarh migrated to Pakistan.

The princely state of the Nizam of Hyderabad was still to make its mind whether it should merge with India or Pakistan or remain as an independent country. A *razakar* movement was launched to defend the Nizam's territory to avoid merger with India. In a swift military action, the Indian armed forces entered the Nizam's dominions and annexed the territory. The Maharaja of Jammu and Kashmir also took several weeks to decide whether he should merge with India or retain its independent status. Meanwhile, armed Pakistani tribesmen overran the Kashmiri establishment. The Maharaja decided to merge with India under pressure. His dilly-dallying tactics resulted in the loss of one-third of his territory. Pakistan was a major gainer in that sense. A line of armistice (line of control) was mutually agreed upon but it became a bone of contention between the two nations.

There are several landmarks in the history of Pakistan, which challenged its identity, relevance and viability as a separate nation. First, the death of Mohammad Ali Jinah, the founder of Pakistan, in 1948 made the state of Pakistan unstable. The top leadership was eliminated. Jinah's right hand man, Nawabzadah Liaqat Ali Khan, the first prime minister of Pakistan, was assassinated in 1951. Thus the two stalwarts of the idea of Pakistan movement were removed from the scene in just three years. The scramble for the top position of the prime minister further ruined the prospects of Pakistan emerging as a viable state. Prime ministers and governments changed time and again. It is noteworthy that Pakistan consisted of two wings. The Eastern Wing was marginalised by the leaders of the Western Wing. The general elections of 1954 could not produce favourable results as the struggle for power resulted in East-West confrontation. Muslim League lost its hold on the nation and a compromise candidate, Iskander Mirza, became the new Governor-General. During his regime a new constitution was adopted, which declared Pakistan as an Islamic Republic with Iskander Mirza assuming the office of the president in 1956. However, it was a provisional arrangement.

Suhrawardy, a leader of the Awami League of East Pakistan, was elected as prime minister. But then he contested elections as a candidate for presidency. He lost the elections. He worked for an abiding East-West alliance. However, President Mirza forced him to resign. A coalition government was formed but it lasted only for two months. In 1958 Pakistan again tried to change over to parliamentary democracy. But Mirza dismissed the government in October 1958. He dissolved the National Assembly. General Mohammed Ayub Khan, the Commander-in-Chief of the armed forces, was appointed as Chief Martial Law Administrator. Within twenty days, General Ayub forced the president to resign. He himself assumed the office of the president. General Ayub promulgated a new constitution in 1962. This was the sixth constitution in the fifteen years history of Pakistan. In March 1969, Ayub resigned and handed over power to General Yahayaha Khan. A martial law regime was established. In 1970, Sheikh Mujib-ur-Rehman, leader of the Awami League, demanded autonomy for the

East Wing. In December the same year, General Yahayaha Khan ordered new elections. Mujib won with a landslide victory. Bhutto's party, Pakistan Peoples Party (PPP), emerged as the largest party in the West Wing but adamant General Yahayaha Khan postponed the convening of the National Assembly. Mujib announced an independent government in East Pakistan. General Yahayaha Khan–Mujib parlays in Dacca, the capital city of the East Wing, failed. Mujib was arrested and taken to West Pakistan for trial. Pakistan army crushed opposition in East Pakistan with a heavy hand. The army unleashed a reign of terror. India intervened on December 3, 1971. This action on the part of India converted into a full-fledged war between the two countries. The Awami League leaders sought refuge in Calcutta. The League declared a government in exile. Under these conditions the Pakistan army found it difficult to win and the entire force comprising 93,000 troops surrendered. General Yahayaha Khan handed over power to Zulfikar Ali Bhutto. After the emergence of an independent nation, Bangladesh National Assembly adopted country's fifth constitution in January 1973. Bhutto who became the prime minister of the West Wing recognised Bangladesh as an independent state. For his part of Pakistan he adopted socialism as the objective of the nation. Another *coup d'état* on April 4, 1977 brought Bhutto down. He was accused of rigging the elections. General Zia-ul-Haq, leader of this *coup d'état*, ordered the arrest of Bhutto, who was later hanged for an allegedly political murder. A year later in 1978, General Zia assumed the office of the president. He amended the constitution and all political parties were banned by March 24, 1981. In December 1984, a referendum was held which endorsed Zia's imposition of the Islamic Law. Fresh parliamentary elections were held and martial law was lifted. In May 1988, Zia dissolved the government and constituted a new government. Three months later, Zia died in an air crash. The office of the presidency was given to Ghulam Ishaq Khan, a civil servant. Fresh elections were held in November 1988. The PPP won the elections with Benazir Bhutto becoming the new prime minister. In August 1990, the president dismissed Benazir Bhutto's government and ordered fresh elections in October 1990. Benazir Bhutto lost the elections. Nawaz Sharif, leader of the Pakistan Muslim League (PML), won the elections with a landslide victory. He became the prime minister and Benazir Bhutto was elected as the leader of the opposition. However, in 1993 Sharif resigned from office and Benazir Bhutto was elected Prime Minister again. In November 1996, President Leghari dismissed Benazir Bhutto's government and elections were held in 1997. In mid-May 1998, India tested a nuclear device at Pokhran and in the end of May, Pakistan tested a similar nuclear device. These developments were indicators of a restless polity. The two nations of India and Pakistan frightened each other by testing their nuclear devices. Both of them asserted that the device would not be used in any war with each other. The two leaders of the government assured the world that the weapon is a deterrent and not a part of the race for arms in South Asia. In 1999, General Pervez Musharraf assumed executive powers, and subsequently named himself President.

Table 7.2: Pakistan: Evolution of Polity

Time Period	*Political Events*
August 14, 1947	New nation is born but partition came with violence; 3.5 million people (Sikhs and Hindus) migrated to India; 5 million Muslims migrated to Pakistan.
September 1947	Junagarh declares merger with Pakistan; India annexes the territory by using force.
October 1947	Pakistan's tribesmen overrun the Kashmiri state. The maharaja joins India under pressure, loses one-third of territory to Pakistan.
1948	Indian army moves in the Nizam's territory; Annexes the territory using massive force.
1948	M.A. Jinah dies.
1951	Liaqat Ali Khan, first P.M. assassinated; Nizamuddin becomes PM.
1953	Mohammad Ali Bogra captures power by replacing Nizamuddin.
1954	General elections; Muslim League routed in the East Wing; Governor-General dissolves the Constitution Assembly.
1955	New Assembly meets; Dominance of Muslim League is over; Mohammad Ali Bogra replaces Chaudhary Mohammad Ali (West Pakistan); Iskandar Mirza new Governor-General.
1956	A new constitution becomes effective; Declares Pakistan as an Islamic Republic; Mirza elected as provisional President.
September 1956	Huseyn Shaheed Suhrawardy, leader of the Awami League of East Pakistan elected as PM; East-West alliance was planned by Suhrawardy; As a presidency candidate he did not succeed; President Mirza forced him to resign; Ismail Ibrahim Chundrigar heads a coalition government which lasted for 2 months.
1958	Fall up with parliamentary democracy, Mirza dismisses government; Dissolves National Assembly; Commander-in-Chief General Mohammad Ayub Khan supports him; Is appointed Chief Marshal Law Administrator.
October 27, 1958	Twenty days later Ayub forces the president to resign; He himself assumes the office of presidency.
1962	Ayub promulgated a new constitution; He names it basic democratic system
1965	India and Pakistan involved in another war.
1966	Soviet government invites the two sides to come to an agreement; They sign (in January) an agreement known as the Tashkant agreement; The two sides agree to withdraw forces to pre-war positions.
March 1969	Ayub resigns; Hands over the power to General Agha Mohammad Yahayaha Khan (in March); Yahayaha Khan assumes the presidential office; He imposes Martial Law.
1970	Yahayaha Khan issues an ordinance against monopolies; Sheikh Mujib-ur-Rehman, leader of the Awami League demands autonomy for the East Wing; He suggests a federal government to deal with defence and foreign affairs; He also demands different currencies for the two Wings of Pakistan—East and West Wings.
December 1970	General Yahayaha Khan orders fresh elections; Mujib wins a landslide victory in East Pakistan; Bhutto's party PPP emerges as the largest party in the West Wing.

(Contd.)

(*Contd.*)

Time Period	*Political Events*
March 1971	Yahayaha Khan postpones convening the National Assembly; Mujib is dissatisfied; Establishes an independent government in East Pakistan; Yahayaha Khan opens negotiations with Mujib in Dhaka; Not successful; Mujib under arrest; Taken to West Pakistan for trial; Pakistan army crushes Mujib's civilian followers; They demand independence; Heavy casualties during military operations; India claims 10 million Bengali refugees have crossed the border; The Awami League leaders take refuge in Calcutta; Declare a government in exile.
December 3, 1971	India intervenes.
December 20, 1971	Pakistani army surrenders to the Indian army; Yahayaha Khan hands over power to Z.A. Bhutto.
1972	Independent Bangladesh comes into being.
January, 1973	National Assembly adopts country's fifth Constitution.
1974	Bhutto recognises Bangladesh; Bhutto the new prime minister; Fazal Elahi Chaudhary new president; Bhutto introduces reforms; Feudal class and religious leaders oppose Bhutto's socialism.
March 1977	Nine opposition parties unite in the Pakistan National Alliance (PNA); Decide to field candidates against Bhutto's PPP; PNA mass demonstrations accuse Bhutto of rigging the elections; Boycott Provincial elections.
April 4, 1977	General Mohammad Zia-ul-Haq leads a *coup d'état*; Removes Bhutto; Bhutto imprisoned.
September 1978	General Zia assumes presidency; Establishes *Shariat* as the law of the land.
1979	Constitution (1973) is amended; Bhutto is hanged.
March 24, 1981	Political parties banned.
December 1984	Referendum endorses Zia's Islamic Law; His presidency extended.
February 1985	Fresh parliamentary election the same year; Martial Law lifted.
May 1988	Zia dissolves the government; Constitutes new government; 3 months later he dies in an air crash; Ghulam Ishaq Khan, a civil servant becomes new president.
November 1988	PPP wins the fresh elections; Benazir Bhutto new prime minister.
August 1990	President Ishaq Khan dismisses Ms. Bhutto's government.
October 1990	Bhutto loses the elections; Nawaz Sharif of the Islamic Democratic Alliance (IDA) becomes the new prime minister. Ms. Bhutto leader of the opposition.
1993	Sharif resigns under pressure; Ms. Bhutto becomes prime minister again.
November 1993	Farooq Leghari new president.
September 1996	Murtaza Bhutto killed in a gun battle in Karachi.
November 1996	Leghari dismisses Bhutto's government; Malik Meraj Khalid caretaker prime minister.
1997	Fresh elections to the National Assembly; Nawaz Sharif leader of the PML.
February 1997	Nawaz Sharif wins landslide victory.
Mid-May 1998	India tests a nuclear device at Pokhran.

(*Contd.*)

(Contd.)

Time Period	Political Events
End of May 1998	Pakistan conducts the nuclear test in Baluchistan.
October 1998	Pervez Musharraf appointed Chief of Army; Relations between India and Pakistan deteriorate.
1999	Benazir Bhutto convicted; Sharif overthrown in coup led by General Musharraf.
2000	Sharif is convicted of hijacking and sentenced to life imprisonment; goes into exile in Saudi Arabia.
2001	Musharraf names himself President.

Bangladesh

There is evidence to show that democracy, secularism and national reconstruction are priority areas and not mere slogans! During the thirty-five years of its history, Bangladesh has passed through inexplicable phases of 'military *coup d'états*': two assassinations—Mujib (August 1975); Zia-ur-Rehman (May 1981); 3 *coup d'états*: first, August 15, 1975; second, 1977; third, March 1982; Parliamentary elections: 1979; November 1981; May 1986; October 1986; February 1991; and February 1996. These developments show how unstable the state of Bangladesh is.

The newly-created nation of Bangladesh was not lagging behind in its political drama after becoming independent in December 1971. A new constitution was adopted in 1975. In a *coup d'état* in August 1975, Mujib-ur-Rehman was assassinated. A military leader came to power. In November the same year, Mushtaq Ahmad was ousted from power. In his place Abu Sadat M. Sayem assumed power. But Sayem was forced to resign and General Zia-ur-Rehman became the president in 1977. In 1979, martial law was lifted after the elections to the parliament. In May 1981, General Zia-ur-Rehman was assassinated. In May 1981, another election was held and Abd-us-Sattar was elected as the president. He was removed from power by General Ershad in a military *coup* in March 1982, who took over as president in December 1983. In November 1986, martial law was lifted but devastating floods of September 1988 destabilised the government. Ershad was forced to resign and was convicted and imprisoned. In February 1991, elections were held again and this time Khalida Zia, widow of the assassinated president Ziaur Rehman, assumed the post of the prime minister. However, in April 1991 a cyclone hit the country leaving 70,000 people dead and it left everything topsy-turvy. Elections were held again in February 1996, in which the Bangladesh National Party (BNP) swept the polls. In June 1996, Hasina Wajid was appointed as the new prime minister.

Table 7.3: Bangladesh: Evolution of Polity

Time Period	*Political Events*
March 1971	Bangladesh constituted.
December 1971	Independence assured.
1975	New Constitution: Mujib assumes the office of presidency
August 1975	Mujib assassinated through *coup d'état*; Military comes to power.
November 1975	K. Mushtaq Ahmad ousted from power; Abu Sadat M. Sayem comes to power.
1977	Sayem resigns; General Zia-ur-Rehman assumes presidency.
1979	Martial Law lifted after parliamentary elections.
May 1981	General Zia-ur-Rehman assassinated.
May 1981	New President, Abd-us-Sattar takes charge.
March 1982	Another military coup; General Ershad becomes chief martial law administrator.
February 1983	Demonstrations against the military regime.
December 1983	Ershad takes over as President.
November 1986	Martial Law lifted.
September 1988	Devastating floods.
December 1990	Ershad resigns; Convicted and imprisoned.
February 1991	BNP wins elections; Khalida Zia assumes the post of prime minister.
April 1991	Cyclone hits the country; 70,000 people die.
February 1996	BNP sweeps the polls.
June 1996	Hasina Wajid becomes new prime minister.
December 1997	Peace treaties signed with tribal leaders to control insurgency.
January 2007	Caretaker government appointed; Paving the way for free and fair elections to be held before end of 2008.

Nepal

The recent history shows that the country passed through several phases, which were marked by violence and civil strife. At one stage the king of Nepal was forced to run away from the scene. India offered him an asylum temporarily. The prime minister of Nepal in response to the Indian prime minister Nehru's appeal agreed to allow the king to return to his kingdom on 18 February 1951. On November 16 the same year, prime minister Rana was removed. In his place a congress party leader, M.P. Koirala became the next prime minister. Despite the opposition of the old feudal regime, the king convened a meeting of the advisory council in July 1952. In 1955 the king died and his son, Mahindra Bir Bikram succeeded as the new king. In 1959, the first elections were held within the framework of a democratic constitution. B.P. Koirala won the elections and was appointed as prime minister. But in 1962 the king dismissed the government in power. The parliament (*panchayat*) was suspended and a new constitution was proclaimed. In 1970, mass opposition to the policies of the royal family during the people's movement sparked widespread riots. Ten years later in 1980 a referendum was held to ascertain the people's views on the form of government. The people heavily voted in favour of a non-party *panchayat* system. As a follow-up, elections were held in 1981. However, the situation was not under control and elections were held again in 1986. The political circles reacted sharply to the king's actions. The ban on political parties was lifted in 1990. In May 1991, Nepalese Congress party won the elections and Girija Prasad Koirala became prime minister. But the political situation has remained fluid.

Table 7.4: Nepal: Evolution of Polity

Time Period	*Political Events*
1324 A.C.	Rajput military castes ousted from north India, overruns Nepal.
1768	Raja's descendents rule over Nepal up to 1768; Country is seized by the invading Gorkhas.
1790	They consolidated their power; Attempt unsuccessful attack on Tibet.
1792	Chinese army defeats Gorkhas; Occupies a part of Nepal, but for a brief period only.
November 1814	British declare war on Nepal; Nepalese forces defeated.
1815	Peace agreement signed with the British; Provide for handing over extensive parts of the terai and the border areas to the British.
1846	Sir Jang Bahadur, a member of the Rana family and a pro-British politician, appointed as PM.
1856	Peace treaty signed with Tibet; Tibet agrees to pay a yearly tribute.
1857	Nepal co-opted with the British.
1923	The British re-affirm the independence of Nepal.
November 7, 1950	Maharaja Mohan Shamsher Rana removes the King (Bir Bikram Tribhuvana), who flees to India.
January 8, 1951	Prime Minister Rana responds to Nehru's appeal to allow the king to return to Nepal.
February 18, 1951	King returns to Nepal.
November 16, 1951	Prime Minister Rana removed; Nepalese Congress Party leader, Matrika Prasad Koirala becomes PM.
July 4, 1952	King convenes Advisory Council meeting in the midst of opposition by the old feudal regime.
1955	King Tribhuvana dies; His son, Mahindra Bir Bikram succeeds as the new king.
1959	First elections; Overwhelming support to the Congress Party; B.P. Koirala becomes new PM.
1962	King dismisses the government, suspends panchayat; A new constitution proclaimed.
1970	People oppose the royalty; Widespread riots.
1980	Referendum to ascertain the form of government; People's choice was to retain non-party *panchayat* system.
1981	Elections held.
1986	Elections held again.
1990	Thirty-year ban on political parties lifted.
May 1991	Nepalese Congress Party wins the elections and Girija Prasad Koirala becomes PM.
Since 1994	Political situation remains fluid.

Sri Lanka

The island nation of Sri Lanka has a long history. According to historical records, Vijaya, a Hindu prince from northeast India, conquered the island in 504 B.C. He called his new realm as Sinhala and encouraged Tamils to settle in the country. As early as third century B.C., Buddhism was adopted as the national religion. The Tamil kings controlled the territory from late third century A.C. to the middle of the twelfth century A.C. The Chinese forces occupied the island between 1408 and 1438 A.C. From 1439 onwards the country was partitioned and ruled by petty kingdoms. In 1517 A.C. Portuguese were allowed to build a fort at Colombo. Subsequently they controlled a large part of the island. However, the conflict between the Portuguese and the Dutch continued unabated and by 1658 the Dutch managed to become the rulers of the country. From 1795, the British naval forces strengthened their hold on the island, which continued for more than a century. By the beginning of the nineteenth century the British occupied the whole island. In 1803, a treaty was signed with the British and the island was ceded to Great Britain. Around the same time native rebellions against the British were intermittently staged in 1817, 1843 and 1848. However, none of these rebellions were successful. But the freedom movement further intensified. In 1931, a new constitution was promulgated according to which Sri Lanka was given autonomy. After a long freedom movement, on February 4, 1948 the island nation won its freedom. One of the frontline leaders of the freedom movement, D.N. Senanayake, became the head of the state. He died in 1952 and his son Dudley Senanayake became the next prime minister. United National Party (UNP) lost the elections held in April 1956. On September 26, 1959 the fourth Prime Minister, Bandara Naike, elected in April 1956, was assassinated. Elections were ordered and the UNP won with overwhelming majority on March 19, 1960 with Dudley Senanayake again becoming the prime minister. On July 20, 1960, general elections were held because the coalition fragmented and Sirimavo Bandara Naike, widow of the slain prime minister, Solomon Bandaranaike, became the new prime minister. Sinhalese was recognised as the official language of the country on December 31, 1960. This move triggered Tamil protests who opposed this decision. The government imposed emergency and the Sinhalese-Tamil relations worsened. But in 1965, she lost the elections and D.S. Senanayake became the next prime minister. Tamil was recognised as the official administrative language in the northern and the western parts of the island. A coalition government headed by Sirimavo Bandara Naike assumed office in 1970. People's Liberation Front, a leftist party, revolted. But these rebellions were suppressed and in September 1971 the senate was abolished and the house of representatives was renamed as National Assembly. The name of Ceylon was changed and it now became the Democratic Socialist Republic of Sri Lanka. A new constitution was adopted in 1972 which ended the country's status as a Commonwealth realm. In 1977, Sirimavo's party was defeated and R. Jayewardene assumed the office of the prime minister for some months before becoming the President. The existing constitution was shelved and a new constitution was adopted. The new constitution provided for an Executive President and as a result R. Jayewardene assumed the office of the president in 1978. R. Premadasa served as the prime minister in the government headed by R. Jayewardene from 1978 to 1989. In 1987, Indian troops entered, Sri Lanka as a peace-keeping force. On Jayewardene's retirement in 1989, R. Premadasa succeeded as the president. New elections were held in 1989 and the UNP retained its majority. From now onwards clashes between the armed forces and the Liberation of Tamil Tigers Elam (LTTE) became common.

Table 7.5: Sri Lanka: Evolution of Polity

Time Period	*Political Events*
504 B.C.	Vijaya (543-504 B.C.), a Hindu prince from northeast India conquers the island, subjugates the Veddas, local aboriginals; His realm known as Sinhala; Encouraged Tamil in-migration.
3rd Century B.C.	Buddhism adopted as the national religion.
Late 3rd Century A.C., to Middle of 12th Century A.C.	Sinhala controlled by Tamil kings; A succession of invaders A.C. to Middle of from south India.
1408-1438 A.C.	Chinese forces occupy the Island.
1439 onwards	Island ruled by petty kingdoms; Island partitioned.
1517	Portuguese build a fort and a trading port at Colombo.
End of the 16th century	Portuguese control a large part of the island.
1638-39	Dutch attacks on Portuguese stronghold; The first of the series of attacks.
1658	Dutch gain control of most of the island; King of Kandy remains independent.
1795	British expedition navy force in the island.
1802-3	British occupy the island excluding the Kandy kingdom; A treaty signed according to which the country was formally ceded to Great Britain.
1817, 1843, 1848	Native rebellions against the British, but unsuccessful; Struggle for freedom intensified.
1931	British promulgated a new constitution; Granting semi-autonomous states in national affairs.
1939-45	The island served as the main base for the allied forces against the Japanese.
February 4, 1948	Island wins freedom, joins the Commonwealth Nations; D.S. Senanayake as PM (leader of the United National Party).
January 1950	Commonwealth Nations assemble at Colombo; Adopt Colombo Plan.
1952	D.N. Senanayake dies; His son, Dudley Senanayake becomes next PM.
1954	Sri Lanka refuses to join SEATO.
December 14, 1955	Republic of Sri Lanka joins UN.
April 1956	UNP loose elections; Bandara Naike, leader of the Sri Lanka Freedom Fighters Party, appointed new PM.
Sep. 26, 1959	PM Bandara Naike shot at, dies on September 26.
March 19, 1960	UNP wins with overwhelming majority.
March 21, 1960	Dudley Senanayake becomes new PM.
July 20, 1960	General elections held; Sri Lanka Freedom Fighters party wins.
July 21, 1960	Sirimavo Bandaranaike, widow of the slain PM, becomes new PM.
December 31, 1960	Sinhalese recognised as the official language of the country.

(*Contd.*)

(Contd.)

Time Period	Political Events
1961	Tamils protest on this; State of emergency declared; Sinhalese-Tamil relations worsen.
1965	D.S. Senanayake wins elections.
January 1966	Tamil recognised as official administrative language in the north and west parts of the island.
1970	A coalition headed by Sirimavo Bandaranaike, assumes office.
March 1971	Marxist-oriented People's Liberation Front leads a revolt.
September 1971	Rebellions suppressed; Senate is abolished; House of Representatives renamed as National Assembly.
May 22, 1972	Country's name changed from Ceylon to Democratic Socialist Republic of Sri Lanka; New constitution adopted.
1977	Sirimavo's government loses the polls; Junius R. Jayewardene appointed as leader of UNP; Assumes the office of Prime Minister.
1978	1972 constitution shelved; New constitution adopted, provides for an executive president; Jayewardene assumes the presidency.
1987	Indian troops enter Sri Lanka to enforce the peace accord between the Sinhalese and Tamils (LTTE).
1989	Jayewardene retires; Ranasinghe Premadasa succeeds.
1989	New elections; UNP retains its majority.
1991-92	Clashes between the armed forces and the LTTE.

India

India before independence was a crown colony of Britain. The people were involved in a peaceful struggle for independence on the lines guided by Mahatma Gandhi. The political developments during this period may be divided into several phases: Phase I: from 1920 to 1930; Phase II: from 1930 to 1942; Phase III: from 1942 to 1947.

The first phase in the struggle for independence was marked by a widespread upsurge in public thinking and expressed in print media and public demonstrations. This was the period of churning of ideas and action programmes. Public discussions were focused on the strategy to be adopted to sharpen the cutting edge of the struggle for independence. This was a period when *Khilafat* Movement under the leadership of Ali brothers (Mohammad Ali and Shaukat Ali) was in full swing. Gradually, the movement converted itself into a rigorous freedom struggle. It was also articulated as a non-cooperation movement. During the same period the country witnessed several upheavals. The British invited a select group of Indian leaders to a Round Table Conference held in London in 1930. The purpose of this Conference was to focus on the political future of the country and the people, including the princes. The delegates representing different sections of the Indian people attended the Conference. Mahatma Gandhi represented the Congress Party at the Conference. Those who attended the Conference represented different shades of opinion drawn from the leadership of political parties. There was also a strong contingent of the representatives of the Indian princes. Interestingly, the

Agha Khan who never stayed in India and was never anti-imperialist also attended. Nehru, however, abstained from participation. Before proceeding to the Conference, Gandhi categorically stated that the Congress was interested in full freedom only. Paradoxically, the Agha Khan supposedly became the leader of the Indian delegation. The British were not very clear in their mind on the issue of full freedom for India. This is evident from the fact that Nehru was convinced that nothing was going to happen at the Conference and that desirable results could not come out of such gatherings.

Box 7.3: The Round Table Conference, 1930

The participants of the Round Tabie Conference represented groups of vested interests in India who were tied up with British imperialism. Nehru was highly critical of the way in which the delegates to the Conference were invited. They were thorough reactionaries. Nehru called this Conference as a useless assembly. The vested interests in his words were imperialist, feudal, financial, industrial, religious and communal. The British delegation consisted of elements of the British Conservative Party. He described the proceedings of the Conference as the mutual squabbles varied by feasting and mutual admiration. It was all jobbery—big jobs, little jobs, jobs and seats for the Hindus, for the Muslims, for the Sikhs, for the Anglo-Indians and for the Europeans; but all jobs for the upper classes, the masses had no look-in (Nehru 1962: 293).

At this stage of growing uncertainty the National Congress Party had no clarity about the suitable strategy to be employed to bring the colonial government to an end and replace it by a viable Indian government. The Party was demanding *swaraj*. But the road to this goal was uncertain. By 1935, the authorities of the Raj introduced a Bill, which came to be known as 1935 Act of India. The Act provided for transfer of power to the local parties. Elections were held in 1937 to elect the members of legislative assembly and the state legislatures. The Congress Party won the elections with overwhelming majority. In the meantime, the Second World War diverted the attention of the Raj authorities to the defence of the country. The Congress leadership, however, launched a non-cooperation movement. When the world was involved in a war against fascism, the Congress Party found the most opportune time for launching a Quit India Movement. In view of the fact that the popular movement had a potential of becoming a threat to the British authority, the senior leadership of the Congress Party was arrested and sent to jail. Many leaders were taken to the Ahmad Nagar Fort. Gandhi himself was lodged in the Agha Khan palace in Pune. With the removal of frontline leadership, the British controlled the situation effectively. No newspapers were available to the internees of the Ahmad Nagar Fort. In the beginning even the writing paper was not supplied. Seen in this context several questions arise and no definite answers are available. They are listed below:

(a) Whether Indian leadership was concerned with the overall global problem of German fascism under Hitler with Japanese as an ally. The battle was going to be a turning point in the history of the world. The battle was between democracy and fascism.

(b) What role did the Congress Party play in launching the Quit India movement? Why did they not think of deferring the popular struggle for the freedom of India till the war was over?

(c) In the absence of the top leadership, who guided the masses for this struggle?

Nehru and other Congress leaders were released from the jail in June 1945. During these eventful years the Muslim League expanded its base among the Muslims because it found the ground open. The League intensified its demand for a separate homeland for the Muslims of the subcontinent. The Congress Party in 1945 was unprepared for this transfer of power. The British government issued a white paper, which was placed in the Lower House of the British parliament. In the spring of 1946, Cripps Mission visited India. Cripps came to India for a dialogue with the leaders of the different political parties. The Cripps Mission concentrated on three main issues. These parlays were held for discussing the political future of the country. The representatives of the princely states participated in the wide-ranging negotiations on their own future. From March 1946 to August 1947, a period of 17 months elapsed. The state of confusion was worst confounded. When Wavell, the new Viceroy, came to India, he announced the British decision of appointing an emergency care-taker government. Hectic discussions started among the different sections of public opinion including the Muslim League. After the initial refusal, Muslim League agreed to participate in this interim government. The political situation was on the verge of anarchy, as the date of the 'lapse of paramouncy' was coming closer. The political leadership was further confused. In February 1947, British Prime Minister Attlee announced that his government would relinquish power by June 30, 1948. Following this announcement political tension mounted in India. There was an apprehension and fear on the possibility of a disastrous Hindu-Muslim civil war. In the meantime, in March 1947, Lord Mountbatten was appointed as the new Viceroy of India. He recommended immediate partition of India. In July the same year, a Bill incorporating Mountbatten's recommendations was introduced in the British parliament. The Bill was supported and approved unanimously by both the Houses of British Parliament. This Bill was termed as Indian Independence Act to be enforced by August 1947. Those 745 days were critical to the political future of the subcontinent. The country was burning and the communal riots spread like a forest fire in many parts of the subcontinent. On 14 August, 1947, India and Pakistan were established as independent and autonomous dominions. All Muslim majority districts were going to become the new state of Pakistan. Both Punjab and Bengal were partitioned as a result of the provisions of the Indian Independence Act. Hindu majority districts remained with India. Most of the 562 princely states were given the choice to join either India or Pakistan. They were also given the option of remaining independent and preserving their individual status. A cabinet was formed with Nehru as the prime minister.

The two princely states—Kashmir and Nizam Hyderabad—were still indecisive. Meanwhile the so-called tribals having their origins in West Pakistan and in full support of the Pakistan militia entered the state of Kashmir and moved into the Valley close to Srinagar. The then Maharaja of Kashmir, Hari Singh, was panicky and agreed to merge with India. The defence of the territories of the Maharaja was transferred to India. In response to this offer India despatched a strong contingent of army and paramilitary forces.

The Kashmir issue was taken to the United Nations where long discussions took place on the future of Kashmir. On October 4, 1947 the United Nations announced a plan for the withdrawal of the combat troops from the state of Kashmir. Both the dominions disagreed with the idea of withdrawal of troops. However, the UN plan was very clear that the Kashmiris would decide their future through a plebiscite after the withdrawal of the armed forces. With this view in mind, Chester W. Nimitz was appointed as administrator of the Kashmir plebiscite. This was the beginning of the Kashmir tangle between the two dominions. No viable solution has been found till today. A Line of Actual Control (LOC) was agreed between India and Pakistan. In the Deccan, the Nizam of Hyderabad delayed the decision to merge with India. On September 17, 1949 Indian forces occupied the territory of the Nizam of Hyderabad.

On January 26, 1950, India became a republic and adopted its new constitution with 395 Articles, 12 Schedules and 83 Amendments. According to the new constitution Dr. Rajendra Prasad was elected to the office of the presidency. The Constituent Assembly reconstituted itself as a provisional Parliament. On March 1, 1952 general elections were announced and subsequently held. As expected the Indian National Congress won with a thundering majority: 364 seats out of 489 constituencies. The party also won the elections for the legislative assemblies in all but two of the constituent states. Second general elections were held in February-March 1957. Congress Party won 366 of the 494 seats. Communists won 29 seats and became the largest opposition party. The Communists also emerged as a leading party in the state of Kerala. The two highest offices of the prime minister and the president were retained as such even after the second general elections.

The real test came when Nehru died in May 1964. The Congress Party decided to appoint Lal Bahadur Shastri as the next prime minister. Meanwhile India and Pakistan were at war on the Kashmir issue. Leaders of the two countries were invited to Tashkent for talks where Kosygin brought the two warring sides for peace talks. The night after the so-called Tashkent Agreement was signed, Shastri died. In his place, Indira Gandhi became the next prime minister. The country's politics underwent a change in June 1975 when a verdict of the Allahabad High Court questioned Indira Gandhi's role in 1971 elections. Indira Gandhi had no other alternative but to announce the national state of emergency. Under pressure from opposition parties, Indira Gandhi announced general elections in 1977. This was the first occasion since 1952 when the Congress Party failed to win a majority in parliament. A new cluster of political parties called the Janata Party emerged as the largest party. Sooner than later, the Janata Party government lost its majority in the parliament. Prime Minister Morarji Desai resigned in 1979. Elections were held in January 1980. Congress I (Indira) won by a large margin. Indira Gandhi assumed the office of the prime minister on January 15 the same year. Her son, Rajiv Gandhi won from his constituency in Amethi. The Sikh movement for a separate state of Khalistan alerted the Central Government. The troops were ordered to take charge of the Amritsar Sikh temple. They stormed the temple, sealed off the shrine; hundreds of Sikhs were killed during the encounter. By the end of June 1984 the troops were called back. But Indira Gandhi invited the displeasure of the Sikh community. She was shot at and killed by her own bodyguards at her official residence on October 31, 1984. Immediately after that, Rajiv Gandhi was sworn in as the next prime minister. Another general election was held in December 1984 and the Congress Party came back to power under the leadership of Rajiv Gandhi. The elections were held again in November 1989 and the Congress (I) lost its majority in the parliament. V.P. Singh of the Janata Dal was sworn as the prime minister. But within a year, in 1990, the split in Janata party resulted in its displacement from power. Chandra Shekhar was elected as the new prime minister. But in a short period of about a year, Shekhar government collapsed. New elections were held in June 1991 but during the election campaign Rajiv Gandhi was assassinated in May. Narsimha Rao, an old Congressman, joined as prime minister. Another election was held in May 1996. Rao's Congress Party was defeated which paved the way for the Bhartiya Janata Party (BJP) to assume power at the centre. Its leader, Atal Bihari Vajpayee, was sworn in as the prime minister. But a few months later his party lost majority support and Devi Gowda became the next prime minister heading a coalition government of 13 political parties. But it was no smooth sailing and the government faced another serious crisis. The Parliament was dissolved in December 1996 and I.K. Gujral was appointed as the head of the caretaker government. Another general election held in February 1998 brought the BJP to power. Vajpayee again headed a coalition government consisting of 23 political parties in March 1998. Atal Bihari Vajpayee's government completed its five-year term. In a new political scenario, Congress (I) came back to power. A coalition government came to power with the support of Congress (I), National Congress Party (NCP) and the left parties. Congress (I) nominee, Manmohan Singh assumed the office of the prime minister in May 2004.

This review of the recent political history of Indian democracy is not very encouraging. Coalition governments formed with the help of no less than two-dozen political parties cannot be taken as the symptom of the success of democracy. The nation has to think about it. In our view, the multiplicity of the political parties is the main source of trouble. A liberal democratic government should take care of the fact that the political parties should constitute themselves into fronts, such as the National Democratic Alliance (NDA) and the United Progressive

Alliance (UPA). This arrangement should be made before the elections to the Lok Sabha but not after the elections. The inevitable conclusion is that Indian democracy is a fragile structure even after sixty years of independence. This is the time when political leadership should come together to constitute fronts. The leadership should evaluate its past history and come to a platform. Perhaps the elections can be more viable if blocks are formed before the elections not after the elections. They should adopt a minimum common programme and go to the polls with a single manifesto. This may reduce the prospects of manoeuvring and manipulation of the elected members of the parliament during the post-election period (Table 7.6).

Table 7.6: India: Evolution of Polity

Time Period	*Political Events*
June 1945	Nehru released from jail; British government issues a white paper on the situation in India.
2nd half of 1945	A new wave of anti-British riots throughout the subcontinent.
Spring 1946	Cripps makes another attempt to negotiate with the Indian parties on the future of India; British Viceroy, Wavell announces an emergency care taker government.
October 1946	Muslim League agrees to participate in the new government.
November-December 1946	Political situation on the verge of anarchy.
February 1947	British prime minister Attlee announces his decision to relinquish power by June 30, 1948; Political tension mounts all over the subcontinent following this announcement; Possibilities of a disastrous Hindu-Muslim civil war.
March 1947	Lord Mountbatten becomes new Viceroy of India; Recommends immediate partition of India.
July 1947	A Bill incorporating Mountbatten's recommendations placed in the British parliament; The Bill gets unanimous approval of both Houses of parliament.
August 1947	India and Pakistan established as independent Dominions; Muslim majority areas to be annexed to Pakistan, Hindu majority areas remain with India; Rulers of 562 princely states merge with India or Pakistan; Hyderabad, Kashmir and Junagarh undecided; Confusion prevails; Nehru appointed as Prime Minister.
October 4, 1947	UN plan for the withdrawal of the combat troops from Kashmir.
March 1949	Chester Nimitz appointed as the administrator of the Kashmir plebiscite scheduled for 1949
July 1949	The Indo-Pak conflict comes to an end; The actual Line of Control between India and Pakistan in Kashmir is agreed upon.
September 17, 1949	Indian forces enter Hyderabad state and Nizam's Dominion is annexed to India.
January 26, 1950	India adopts the new constitution.
January 1950	Dr. Rajendra Prasad elected as the first president of India in accordance with the new constitution.

(Contd.)

(Contd.)

Time Period	Political Events
January 26, 1950	India becomes a Republic; The Constituent Assembly reconstitutes itself as a provisional parliament.
March 1, 1952	General elections announced; Indian National Congress wins 364 of 489 seats; Congress also wins in all but 2 of the constituent states.
January 26, 1957	Kashmir is declared as an integral part of the Indian Republic; Protests, riots and burning effigies of Nehru in Pakistan.
February-March 1957	Second general elections; Congress wins 366 of 494 seats of the National parliament; Communists win 29 seats; Become the largest opposition party in the parliament; Communists also gain control over the state of Kerala; After the elections, no change in prime minister and the president; The same leaders continue.
May 1960	The state of Bombay divided into two states of Maharashtra and Gujarat.
May 27, 1964	Nehru dies; Shastri becomes next prime minister.
1966	Shastri attends the Indo-Pakistan meeting in Tashkent; Dies after signing Tashkent declaration; Indira Gandhi becomes the next prime minister.
May 18, 1974	India implodes its first nuclear device.
June 1975	Indira Gandhi is convicted of undesirable practices during 1971 election; She declares national state of emergency.
1977	General elections held; Indira Gandhi loses her seat in parliament; Congress loses its hold for the first time since 1952; Congress fails to win the majority in parliament; Janata Party emerges as largest party.
1979	Janata Party government loses parliamentary majority; Desai resigns.
January 1980	Elections held; Gandhi's Congress Party wins a major victory; Indira Gandhi's son, Sanjay dies in a plane crash; Her elder son, Rajiv replaces him in parliament after wining the elections.
June 2, 1984	Amritsar Sikh temple sealed off by troops; The armed forces occupy the temple; Hundreds of Sikhs were killed during protest agitation.
End of June 1984	Troops are withdrawn.
October 31, 1984	Indira Gandhi shot and killed by her own guards; Her son Rajiv appointed as the next PM.
December 1984	General elections held; Congress party comes back to power under Rajiv Gandhi's leadership.
November 1989	Congress loses its parliamentary majority; V.P. Singh becomes next PM.
1990	Janata party splits; Minority government collapses; Chandra Shekhar joins as next prime minister.
May/June 1991	Shekhar government steps down; New elections, Rajiv assassinated during the campaign; Narasimha Rao leads the Congress party
May 1996	General elections held; Rao defeated, loses parliament majority; BJP emerges as the largest party; Within a few months another coalition government with Deve Gowda as the PM, heading a 13 party government.
December 1996	Government in trouble; Lok Sabha dissolved; Gujral appointed as head of the care-taker government.
February 1998	General elections; BJP emerges as the largest party; Another coalition government comprising 23 political parties; Vajpayee sworn in as PM in March 1998.
May 2004	Manmohan Singh assumes office of Prime Minister.

A Resume

In the foregoing chapter an attempt has been made to trace the origins of Hindu polity. The political institutions, such as the *panchayat*, the *sabha* and the *samiti* are very old. The people of the subcontinent have a fund of collective wisdom. It is this capability, which has helped tide over the long phases of foreign domination. The two centuries of the Iranian rule was followed by the Greek invasion of the subcontinent and control of vast territory. The Greeks ruled over the region for another 200 years. The subcontinent presents a unique example of how the different dynasties of kings rose and fell fighting among themselves. The people were well-versed in the art of living under unhelpful regimes, such as those of the Afghans and the Tughlaks. In between came the Lodhis and the Mughals. By 1757 the British were poised to take charge of the entire subcontinent. Kings and emperors came, imposed their political tradition while the people isolated themselves in rural areas carrying their own indigenous traditions of self-governance and peaceful coexistence. The famous Urdu poet, Firaq Gorakhpuri, once stated that the people of the northern region of the subcontinent survived the supreme authority of the Muslim Sultans and dynasties. In his view, they led their lives under the shadow of the shining sun of the Muslim *sultans*. Tulsidas's *Ramcharit Manas* was like a banyan tree under the shade of which the Hindus and the common people passed the difficult times unaffected and unconcerned by the rise and fall of empires. The *Ramayana* offered them solace. One may conclude that they forgot the present and lived in the glorious past. The people lived in their own small worlds in a state of resignation. There were multiple ethnic groups and multiple polities. However, by the time the British took charge of the Indian affairs, the national leaders from Tilak and Gandhi to Nehru, Maulana Azad, Baba Saheb Ambedkar, Jayprakash Narayan, Subhas Chandra Bose and Ram Manohar Lohia galvanised them to face the British ascendancy. They encouraged them to struggle for an honourable status after the *swaraj*. Non-violence was there in their blood stream and they stood up as one voice of dissent and roused to submit to the over-arching authority of the British. It has been argued that Gandhi's teachings on non-violence were influenced by the philosophy of Buddha who advised not to raise arms against anybody. The two hundred years of the British rule in India is dotted with courageous rebellions and revolts against the ruling power. Millions of Indians from Kashmir to Sri Lanka and North West Frontier Province and Baluchistan to Bengal and Assam stood up to demand freedom from the British rule. Leaders like Khan Abdul Ghaffar Khan, Maulana Mohammad Ali, Maulana Hasrat Mohani, Abul Kalam Azad and Dr. Ansari to mention only a few Muslim leaders were there hand in hand to fight the battle for independence. It would be unfair not to mention their names in the list of freedom fighters. Many of them were ahead of Gandhi in demanding complete independence for India. They suffered long jail terms but did not budge an inch from their position, which was to struggle for *swaraj*. In serious writings on the struggle for independence very often an impression is created that the Muslims were generally indifferent to the struggle for freedom. They were, of course, in a pitiable minority and the masses were swayed over by the Muslim League leadership.

We have tried to portray a picture of the local institutions from the *panchayat* to the *sabha* which followed the egalitarian tradition of democracy. The issues of nation, nationality and nationalism have been discussed to show that our democratic tradition of the people of the subcontinent is as old as its history. One can cite the example of *Janapadas* of ancient India. As we already know, *Janapadas* were People's Republic. There were, of course, kings, and king-emperors but the people were not swayed over by their authority. The rise and fall of empires could not subdue them. A reference has also been made to the origins of the Hindu polity. The framers of the Constitution of India, who visualised India as a federal state, addressed the related controversy of relations between the Centre and the region. Federalism was accepted as the measure to the balance of power between the Centre and the regions (read states). The expectation was that the federal constitution would function satisfactorily. But it did not. The centrifugal force operated strongly on the periphery of the Indian states. The people on the periphery asserted their individuality and demanded a place for them in the national political space. The Tamil movement for self-rule is not relevant now. Naga polity and the Kashmiri demand for a separate status are some of the examples of this phenomenon. The on-going Naga struggle for Greater Nagaland is an example of the people's fluidity in the division of political space. The chapter also presents an outline of the recent political history of the South Asian nations. The recent political happenings in Afghanistan, Pakistan, Bangladesh, Nepal and Sri Lanka suggest that there is nothing short of a drama which is being enacted on the South Asian stage. The rise and fall of political parties and the history of power capture narrates a sickening tale of *coup d'états*, assassinations and change of governments every now and then. So far as India is concerned, democracy has always been a mark of distinction and seems to have been functioning satisfactorily. But the fragmentation of the political space continues unabated. Look at the recent history of government formation at the Centre and States. With two dozen political parties, mostly regional, forming the coalition government is not an easy pill to swallow by those who are devotees of liberal democracy.

The recent political history indicates that the spatial fragmentation of the party following has been a major trend since independence. This is also evident from the history of elections covering the period 1971-1980. The data used for this exercise consist of the spatial pattern of party victories, their relative strength in the Lok Sabha constituencies and from time to time swings in the votes polled by the parties and the consistency in party following. The purpose is to identify the geographic patterning of party domains and to draw conclusions. The emergent patterns in people's voting behaviour reveal political fluidity. The data reveal that the polity is very much amenable to further fragmentation. This confirms that the centrifugal force is operating in the peripheral areas. One observes a genetic relationship between the political behaviour of the people.

No one can ensure that the electoral behaviour of the people has any amount of permanency. People's participation rates differ from constituency to constituency. There is a role of current events that results in massive shifts to the party following. The day-to-day issues, for example, the price of onion or police firing on the masses due to some local, regional or ethno-lingual groups may lead to massive shifts in people's party following. As we already know, Indira Gandhi, the then prime minister, imposed emergency in June 1975. The next elections to the Lok Sabha in 1977 resulted in the defeat of the Congress Party. But when she was assassinated in October 1984, the polity underwent a dramatic change with her son emerging as the leader of the Congress Party. There was a sympathy wave throughout the country and her son, Rajiv Gandhi, was elected and appointed as the next Prime Minister Several conclusions can be drawn:

(a) The party following is genetically related to the local electorates of religious or caste affiliation.
(b) There are dominant castes in every constituency and the election results depend on their party. For example, prominent leaders of the Congress Party, *viz*. Maulana Abul Kalam Azad, Mohsına Kidwai, or Jaffer Sharif contested their elections from the constituencies in which Muslims had a dominant share in electoral politics.
(c) The distribution pattern of castes and sub-castes in different states of India operates as a major factor in the selection of candidates. If a certain constituency is characterised by the population of a particular caste, all political parties are likely to field their candidates of the same caste.

The recent electoral history indicates that there exists a high degree of political fragmentation in the polity. The phenomenon of fragmentation is evident in the multiplicity of political parties as well in the shifting allegiance of the people of different political parties. The first elections held in 1952, proved that the Congress Party had a sweeping victory of two-thirds of the total seats for which the elections were held. Today, the position of the Congress Party is so weak that it was able to constitute a new government only with the support of other regional parties or its own splinter groups, like the NCP. The leftist parties supported this combination but did not join the government. The same was the case with the BJP who could manage a majority with the help of 23 regional or sub-regional parties in 1998. It is not a good sign but nothing can be done. This is the age of coalition governments at the centre as well as the states.

END NOTES AND REFERENCES

1. Chakravarti *raja* or the *raja* of the whole world. 'For the first time we read of great kings whose historicity is certain and some of whose achievements were known and from now on the main lines India's political development are clear'. The Aryan tribes were ruled by the chiefs who bore the title *raja*, a word related to the Latin *rex* (Basham, 1964: 45).
2. The common people, whose affairs were controlled by the *raja*.
3. Something which is beyond the ambit of human effort. The king was supposed to be an incarnation of god.

4. The *raja* was not an absolute monarch, for the government of the tribe was, in part, the responsibility of the tribal councils, the *sabha* and the *samiti* (Basham, 1964: 33).
5. supra Chapter 2.
6. The King Emperor was a sovereign authority and whose territories extended over vast areas of the Indus and Ganga valleys and the north Indian plain.
7. An ethnic group mentioned in Sanskrit and Pali texts.
8. Successfully assimilated (Thapar 1977: 53).
9. Human development implies social development for which the most critical indicators are education, health, housing (Mahbub-ul-Haq, 2002, see p. 63, Human Development Indicators for South Asia, *infra* chapter 6).
10. A system of governance based on the Islamic law. Many provisions of the law have now become obsolete. However, many a time Muslim countries of the world, including South Asian nations, such as Pakistan and Bangladesh have now described themselves as Islamic regimes. This is a doubtful proposition.
11. Interventions from outside lead to the disruption of the agricultural economy. This is because of the fact that the fields are not attended by the people who have fled away from the scene of trouble.
12. See also Minority Rights Group International (MRG, 2001, p. 6).
13. ibid, p. 6.
14. The reference is to Taliban's insistence on Islamic fundamentalism. The way of living is dictated by the clergy who banned un-Islamic things, such as cinema, theatre, T.V., beauty clinics for women and the like.
15. *Coup d'état* means violent and unconstitutional change in government by persons in authority, particularly the armed forces.
16. My apologies to Gunnar Myrdal, who evaluated the development needs of the Asian countries in his famous book the *Asian Drama*.
17. Cultivation of opium and heroin is an important source of income. Paradoxically, the Taliban movement did not impose any conditions on this trade. In fact, it is alleged that they were the main beneficiaries from this illicit trade. Opium is cultivated in the Helmand Valley in southern Afghanistan and one-half of the opium production comes from Pushtun area. Another twenty-five per cent comes from Nang Har in the east.

8

Development

An attempt has been made here to restate the objectives of the regional development theory and practice as based on the experience of the South Asian nations. The issues of sustainable development and the underlying strategic imperatives have also been discussed. The state of the social and political environment within which goals are supposed to be achieved by the states of the subcontinent, have also been analysed.

Development is a process implying a progressive change, which is believed to be irreversible. However, this statement is not always correct. Economic history shows that development can be reversed under certain specific conditions. Whatever stage the development has reached at a given point of time in the history of a country the basic question remains: can it be sustained? In fact, problems of sustainable development have acquired an added dimension as the debate on the paths of development goes on. The problem is further complicated due to globalisation, implying a process of integration of the economies of the local communities and the peripheral regions into the world economic order. Similarly, liberalisation or the economic reforms mean relaxing conditions for free entry of the local and foreign capital, goods and services and the accompanying baggage for changing the climate for so-called investment. It appears that the Indian political economy is trying to adjust itself to the changing conditions. Seen in this context, the question of sustainable development acquires new meaning. The reason is that the context of development-environment relationship is undergoing a qualitative change. This is besides the social ferment in the South Asian society. In fact, it is revolutionising the very process of social and economic change. The old production relations are no more valid. The traditional ethical values are being replaced by a mercantile culture. The impact of the economic reform policies is yet to be seen.

Development is one of those dreams, which have seized the imagination of the political leadership all over the world. The subcontinent is no exception. It is a global phenomenon, although it is more important in the case of developing countries. The very concept of 'developing countries' is as vague as the concept of 'developed countries'. The underlying assumption is that some countries of the world are supposedly 'developed', while others are 'underdeveloped' or 'developing'. The leaders of the underdeveloped countries were not happy with the description of their countries as 'underdeveloped'. They insisted that a better word would be 'developing'. The United Nations (UN) and the countries of the North (read developed) accepted this logic. It was suggested that the developing countries were always developing and their economies were not stagnating as such. However, there were exceptions to this general rule. They were also given a package of new ideas and new strategies to make the people of the developing countries conscious of the fact that despite their lagging state, their approach to development was positive. But the substance of this debate was that these countries would continue to be underdeveloped and would be required to make extraordinary efforts to catch up with the developed countries of the North. The United Nations launched several development projects suited to the needs of the people in the developing countries. International agencies, such as the World Bank, United States Agency for International Development (USAID), and the International Monetary Fund (IMF) took a lead in evolving and implementing suitable strategies for the development of the underdeveloped world. A similar initiative was taken by other funding agencies, such as the Canadian International Development Agency (CIDA) and the Asian Development Bank (ADB). The ideologues of the development concept came by and large from the North. The United Nations Development Programme (UNDP) and the World Bank recommended new strategies. New schemes were introduced with the help of funds flowing from the international agencies. New strategies, which produced results, were tested in the different parts of the world. It may be noted that during half-a-century from 1950 to 2000, the developing countries continued to be 'developing' in the future-imperfect tense. International agencies are supposed to be playing a critical role in the smooth transition of these countries from the stage of 'developing' to the stage of 'developed'. However, the word 'developed' has an inherent conceptual confusion. The international agencies facilitated the transfer of capital, goods, services and technology. The aim was to improve the economic conditions in the lagging countries of the world. The international funds continued to pour in but not without conditionalities. The poverty scenario in these countries did not change much. In fact, it added to their debt burden. As some writers believe, they fell in a debt trap.

Development policies acquired nuances of their own when a distinction was made between sectoral development and regional development, between economic development and human development. It has been observed that the gap between the lagging and the progressive sections of population has widened. The past experience shows that planning efforts could not produce the desired results. In our view, when development came to the developing countries of the South, the poor became poorer and the rich became richer. This impression was formed on the basis of observed ground realities. More and more people were found sliding down the poverty line.[1] The main reason behind these failure stories was the phenomenal growth of population. This unregulated increase in population overshadowed the gains of development. No amount of development schemes can change the grim scenario. The race between development and underdevelopment was bound to be lost. This governance-prosperity-poverty syndrome was never going to disappear. In a nutshell, the nations of the South lost their battle on the demographic front and they were bound to lose it on the economic front. Moreover, the free market economies were not really free. The symptoms of their decline in the economic domain were too obvious. Whatever they did further resulted in more indebtedness.

Let us examine the concept of development. Admittedly, development is a complex concept (Box 8.1, 8.2, 8.3). For some people, development means widening of roads, converting two-lane roads into four-lane roads, construction of the malls (modern department stores), flyovers, and tube railways in metro cities. Others believed that enhancing the purchasing power of the common men and women and improving their quality of life were far more important. The free global markets were more important indicators of progress. There were other dimensions of development and underdevelopment. They may be listed as follows: annual growth rate of population, level of urbanisation, illiteracy rate, health status of the people, infant mortality rate, life expectancy at birth, hospital facilities and the proportion of population having access to safe drinking water. They were far more sensitive indicators of human development. For the people, human development was the crux of the matter. International agencies like the UNDP and the World Bank collected a mass of data on these indicators to apprise themselves with the state of the art and to ensure that the gaps in human development were reduced. The development problems should be located in space/place so that the rural habitat and the rural folk are addressed. The data were collected from all countries of the world. While it is true that these statistics helped the governments and the people in countries all over the Third World, neither UNDP, nor the World Bank or IMF had any command on the status of data, its reliability and consistency. Nor were they in a position to introduce corrective measures. The data thus supplied by the countries of the world put question marks on their accuracy. The data were cleansed, but could not be authenticated. An interesting race between the development strategies and the capability of the implementing agencies in the Third World counties goes on. The most important objective is not development but the political economy of development. Governments in the Third World countries changed over night with the result that the incoming development funds were either not utilised, or diverted to other heads of governmental expenditure. Very often the priorities were changed. As indicated earlier, what was left behind was the burden of international debt. While the number of people living below the poverty line is increasing, the pro-establishment policies failed to bridge the gap between the rich and the poor. Exceptions apart, the planning policies alienated the rural masses as well as the urban poor.

It may be posited that the process of development was neither neutral to social structure nor independent of it. In fact, it created its own social institutions and led to restructuring of all social and economic forces at all levels of production hierarchy. The development experience of the third world societies, including the countries of the subcontinent, has not been encouraging. It is now generally realised that the attainment of political freedom is not sufficient for achieving the goals of economic development, equity and social justice. According to the official estimates some 260 million people of India live below the poverty line. Non-official estimates point out a larger figure, say from 300 to 400 million people. The policies of economic and regional development initiated since independence in 1947 have further accentuated the regional and inter-personal disparities. In fact, instead of bridging the gap between the rich and the poor the hiatus between the two has widened. In the final analysis, the goal of harmonious human development remains a dream for millions of people in the developing world.

Box 8.1: Environment and Development

Development involves a progressive transformation of economy and society. A development path that is sustainable in a physical sense could theoretically be pursued even in a rigid social and political setting. But physical sustainability cannot be secured unless development policies pay attention to such considerations as changes in access to resources and in the distribution of costs and benefits. Even the narrow notion of physical sustainability implies a concern for social equity between generations, a concern that must logically be extended to equity within each generation.... The satisfaction of human needs and aspirations is the major objective of development (World Commission on Environment and Development 1987: 43).

The debate on the question of balanced development is sharply focused on the objective of sustainable development. The question is whether development as a process is transitory or sustainable over time. Sustainable development is a comprehensive term. It implies that the development needs of the present generation of people would be fulfilled without compromising or interfering with the ability of the future generations to meet their own needs. The two concepts that are basic to the idea of sustainable development are: first, the concept of needs, which in the words of World Commission on Environment includes the essential needs of the world's poor wherever they are and whatever stage of development they have reached. The Commission attached the highest priority to this objective. Secondly, the idea that the environment imposes limitations on the human freedom to exploit the natural resources according to the will of the people is crucial. Moreover, technology and social structure play an important role in utilising the environmental potential to the optimal level. What is fundamental is to meet the present and the future needs of the humankind. Nations and the social groups within the nations (ethnic, caste or class formations) are constrained by the fact that they have basically an unequal access to resources and technology. Moreover, social structure differs vastly in different parts of the world. A major objective of development planning is to provide conditions for the fulfilment of the essential needs of the people, such as food, shelter, clothing, health, education and employment. Any strategy for sustainable development will involve restructuring of the economies the world over so as to ensure growth with equity and social justice. Implicit in this argument is the need for transformation of relations of production and redefinition of the role of social structure in this process of transformation. If sustainable development means 'meeting the basic needs of all and extending to all the opportunity to satisfy their aspiration for a better life', then its goals can be achieved through a radical change in the relations of production.

The roots of the dilemma of our generation lie in our traditional conformism to the existing social system. It is the social system, which ensures the overlordship of the high castes, particularly in rural areas. The system has to ensure that the tribal (*jirga*) formations, lagging castes and the deprived and the underprivileged, have their share in the surplus production. While reorganising the social forces of production would be a long-drawn exercise, it may be suggested that a beginning should be made. The agenda for change is based on the following assumptions: (a) the existing living standards, as they are the only basic minimum, are not sustainable in the long run; (b) lavish living is beyond the ecological means and for that matter beyond the economical means of the nations; (c) the growth potential of the regions of the subcontinent has not been realised completely or it has been realised for promoting the needs of a section of population; (d) the principle of sustainability of economic growth involves absolute exploitation of the people, this would mean that high economic growth and high poverty can co-exist; and (e) population size and growth are a crucial link, but more crucial is the rapid demographic growth, which if accompanied by unequal economic growth, would be a potential threat to the very stability of the state.

Keeping in view the logic of sustainable development, it is imperative to: (a) revive growth in regions where the economic growth has been retarded and the surplus generated has not been distributed equitably so as to enable the poor to meet their basic existential needs. Reviving growth also means reducing disparities between social groups, between regions within a nation and between the nations; (b) initiate qualitative change in economic growth which implies change in the content of growth so as to make it less dependent on materials and the energy resources available in nature; (c) create conditions for the fulfilment of the basic human needs.

Box 8.2: Development and Growth

Development is the value concept and as such it is an elusive concept. The definition of development can be the definition of an individual's value preferences. Whereas conventional *laissez faire* economists associate development with growth in gross national product (GNP) and its associated per capita income, welfare economists on the other hand, emphasise values associated with the common good or public welfare. Therefore, the latter would tend to associate development with rising net income accompanied by reduction in unemployment and social inequality. For the practising politician and the common man, development is about modernisation or the acquisition of facilities, services, and infrastructure associated with modernity, such as clean and safe water supply, health facilities, and all-weather roads. The notion of development, therefore, needs a more precise definition if one is to appreciate whether there is any relationship between it and a project of decentralisation (Oyugi 2000: 4).

The present pattern of relationship between the natural environment and economic growth is such that a majority of the people of the world cannot meet the basic needs of food, clothing, shelter, safe drinking water, sanitation, education and health. The conditions are not very encouraging. Millions of people have no idea about the balanced diet. Even if they know, it is not within their means to prevent under-nourishment and starvation; (d) check the population growth in the nations of the world as well as the sections and groups within the nations for maintaining a manageable size of population. Although population growth will continue to be uneven, it will be necessary to strike a balance between population size and natural resources. This objective can be achieved by limiting the excessive use of resources by the industrially developed countries of the world and the consequent deprivation of the poor; and (e) suggest alternative use of resources for checking the prevailing situation which threatens the prospects of resource conservation globally. The pressure on resources can be minimised if alternative uses of available resources are introduced and implemented. This would mean that alternative technologies are the need of the hour for avoiding ecological disasters. The developmental needs of the poor cannot be met unless alternative technologies are discovered. The industrially developed countries should be told to minimise the ecological degradation and the risk involved therein. The survival of the poor, particularly in the developing countries of the world, would depend on the transfer of suitable technology as the first major priority[2] (World Commission on Environment and Development, 1987).

Box 8.3: On Development

A process of becoming and a potential state of being. The achievement of a state of development would enable man to make his own history under conditions of his own choosing. The process of development is the means by which such conditions of human existence might be achieved. They in turn, would necessarily involve man in a productive, crisis-free, and non-exploitative set of relations with nature and in the struggle to remove oppression and exploitation from the relations between men. By these standards, no society in history has ever achieved a state of development and it may be argued that no society has ever engaged in a process of development (Lee 1983).

Social environment, like the natural environment, has several dimensions that make it functionally viable. It enables the people to function as an organic whole. These elements depend on the nation's collective as well as individual social commitment to the objectives of equity, social justice, and equality of opportunity. The desirable social environment for achieving the goals of sustainable development can be created, provided the political system ensures citizen's participation in decision making, irrespective of ethnicity, caste, colour, creed, religion, social standing and political party affiliation of an individual. Further, provided that the economic system ensures equitable distribution of the socially generated economic surplus by allowing free access to technological knowledge and means on a self-reliant and sustained basis. The basic problem would remain intact unless the administrative system commits itself to the goals of harmonious development. It implies that the system should strive towards building an egalitarian society by distancing itself from the rich and the political elites (ruling classes). The system should lend support to the weak, underprivileged and the downtrodden (the oppressed classes), if it has the capacity for self-correction and the redressal of grievances. The condition is that the system respects the obligation to preserve the ecological base for the present and the future needs of the people. If it were not able to resist monopolistic exploitation of natural resources, it would result in the eventual depletion of resources. The prevailing situation is that there are communities to whom the benefits of development have not accrued. Their experience is discouraging. They assert that when development comes they are not the beneficiaries; only industrial waste is transferred to them.

The people in the underdeveloped or the developing parts of the world have no choice under the over-arching economic and political order of the day. Nations can achieve these goals if the international order ceases to be detrimental to the interests of the poor in whatever countries they are found. Sustainable development will remain a far cry if the hegemony of the developed nations of the North goes unchecked and unchallenged. It would remain a dream if the existing relations between the rich and the poor nations were allowed to continue.

Development History

At partition, it was estimated that after exodus from either side of the Indo-Pakistan border there were about thirty-two million Muslims in India and fifteen million non-Muslims in Pakistan. In the paragraphs that follow we will discuss the status of development in India and Pakistan on the eve of partition, taking this as the baseline on August 15, 1947. We will discuss economic impact of the British policies in India. The British policy affected India's agrarian economy, particularly the land relations, tenural relations and agricultural development; industry, factory as well as artisan industry; transport infrastructure, particularly railways, roads and waterways; infrastructure for health and education; structure of the export and import trade; and rural-urban linkages.

Before we go to a detailed presentation of these aspects of the economy it would be necessary to recognise the British contribution to India's development.

British Legacy

The two dominions of India and Pakistan received an impressive heritage of developmental assets and infrastructure from the British. However, the greatest contribution of the British was the integration of the territory and developing an all-India view. It is paradoxical that the concept of India as a state was a gift of the British. As already stated, India was never a unified state ruled by a single authority before the British; it was never a unified state after the British. It is true that they retained hundreds of princely states as adjuncts to the British Empire. They left the choice for the princely states between several options: join India or Pakistan; they were also free to declare their independent and autonomous status. The international boundaries of the British India became the boundaries of the two independent dominions of India and Pakistan.

For the ministers of the first Government of India under the leadership of Pandit Jawaharlal Nehru, it was not a cakewalk to power, even though they inherited the entire governmental super structure. It was not a cakewalk because they reached that stage after walking on fire. It was an irony of history that the transfer of power appeared as a change of hands from white to wheatish; the model of governance did not change. The new government was a continuity of the old. The first Governor General in the person of Lord Mountbatten was an evidence of continuity; the bureaucracy, in appearance and functioning was a continuity. Interestingly, it was the secretarial staff that guided the new government to make a smooth transition from the British to the Indians. The development-related work was inherited from the British Raj to the self-rule (*swaraj*). The Nationalist Government received an edifice for new India's further development, indeed a superstructure for the liberated masses of India. The argument that the withdrawal of the British from power was a landmark in the political history of the nation and the Indian state is half the truth. The other half-truth is that the process of liberation of the people from the foreign rule was an act of magnanimity. All policies/programmes of the British government were, by and large, oriented towards the improvement of the quality of life of the teeming millions of India. The new schemes of development were already there in place before the national government took over.

Successive governments in England and the British parliament kept an eye on the state of the British India, its economic development and schemes of upliftment of the masses from poverty and squalor. There were long debates in the British parliament on the issue of governance and indigenous development in India. This is notwithstanding the fact that development planning left much to be desired. So are the actions of the governments who came to power in India, Pakistan, Nepal and Sri Lanka since independence. Indigenous governments failed on many fronts, just as the governments failed in other countries of the world. Monitoring the acts of commission and omission was the responsibility of the British parliament. The British press and the public men never missed an opportunity to pull up the British administration in India.

There is no gainsaying the fact that the British in India were assisted by the competent Indians in all spheres of work and all spheres of life including administration. The governmental staff included highly competent and skilled secretaries and officers trained in the British tradition. It will be wrong to say that the civil service was all in all pro-British. But the civil servants always sided with the ruling class. It is a fact that the trained secretariat helped the new government functioning without a break from the past. The first nationalist government under the leadership of Nehru started building on the edifice laid down by the British.

Another important British contribution to development was in raising a highly skilled administrative force equipped with an experience of governance and the knowledge of the procedures involved. Thus we can say that the first nationalist government of India started to play on the firm pitch left by the colonial rulers. The British also demonstrated the functioning of the democratic parliamentary tradition with a record of excellence including infrastructure for the new nation. It was this democratic and constitutional functioning that was the hallmark of the British achievements. Examples can be cited to prove this fact. They left behind a central Legislative Assembly House (now called as the Parliament House). It is a bicameral body consisting of an upper house (*Rajya Sabha*) and a lower house (*Lok Sabha*). Its procedure was laid out in the law and the written tradition. The North and South blocks, Viceregal lodge (now called Rashtrapati Bhavan) are the landmarks and a memorial to their monumental work and foresight. Seen in this perspective, the British were the builders of a democratic tradition. There were, however, many areas of public functioning in which the British were not obliged to take interest or action. Identical facilities were created at the provincial headquarters, e.g., assembly houses in Lucknow and in other provincial capital cities, such as Bombay (now Mumbai), Calcutta (now Kolkata) and Madras (now Chennai). There was a competent super structure available for maintaining law and order as well as for the smooth functioning of the governments and for ensuring social security—a police force, para-military force, e.g., Provincial Armed Constabulary (PAC); home guards; an armed force comprising army, navy and the air force. This is besides the point that this force was sometimes also used to defend the British interests in India and abroad. In retrospect it appears that there was nothing unusual about it. Even today it is a common practice that the independent nations of the world send their military force to the troubled countries for peace keeping under the aegis of the United Nations. The three wings of the defence force were given an elaborate system of ranking and promotions. For example, regiments were constituted on ethnic lines—Sikh Regiment, Jat Regiment, Rajputana Rifles, Gorkha Regiment, etc. They established a system of civil defence, an elite police force, a judicial system from the *tehsil*-level *munsif* to the High Court judges; a district judge and a Sessions Court at the district headquarters. The delivery of justice was streamlined with a back up of a legal procedure laid down in the public law and a penal code, IPC (Indian Penal Code). They established a system of criminal procedure for the punishment of deviant elements. Gaols were constructed for the confinement of the accused and the criminals. Every district had a gaol which functioned under a system of administration. An administrative set-up was created for each district comprising a district magistrate or/and a collector/commissioner, a sub-divisional officer, a *tehsildar*, etc. Every *tehsil* had a government treasury; excise, customs, and a tax collection system of land tax (revenue), income tax, sales tax, etc. There was an excise duty to be levied on every item produced. This was done on the basis of a licensing policy. There was a law to cover every activity undertaken by the public and licensed by the government.

This is not to say that there was no law before the British came to India. There was a Mughal law, highly elaborate and comprehensive for the punishment of those who violated the public law and for the transaction of justice. However, the decline of the Mughal Empire after the death of the Mughal Emperor, Aurangzeb in 1707 created a vacuum of power. The empire was dismembered with the result that the new centres of power emerged in the provincial capitals, each *subedar* (provincial governor) declared himself as the *nawab* or the *maharaja*, a symbol of his independent and autonomous status. Naturally, the fragmentation of the empire led to lumpen elements roaming freely and violating public law. Reference may be made to Lord William Bentinck's campaign against the *thugs* (fraudulent elements not much different from highway robbers) who openly looted and made the life of law-respecting citizens a hell.

It is well-known history that the British India time and again suffered from famines due to the erratic nature of the monsoon, the main rain-bearing winds. In order to meet these contingencies a huge network of irrigation canals was developed, particularly in the dry regions of the subcontinent, such as Punjab and Sind where the Satluj, Chenab and Jhelum rivers carried large volumes of water. The river-waters were transferred to the agricultural fields through the canals ensuring irrigation of the crops on a large scale. This brought about a revolutionary change in the cropping pattern. These otherwise dry regions turned into the producers of the best quality of wheat, rice and cotton. Similarly, agricultural situation in the Ganga-Yamuna *doab* improved remarkably as a result of the construction of the Upper and the Lower Ganga canals. These canals stabilised the production of crops, such as wheat and sugarcane. In fact, sugarcane became the basis for the sugar industry. It also supported the indigenous *khandsari* industry.

A huge railway network of thousands of miles/kilometres was laid out. Both broad gauge and meter gauge lines were constructed. The railways connected every nook and corner of the subcontinent. More particularly, the railways linked the port centres like Bombay, Calcutta, Madras and Karachi with their vast hinterland. Besides railways, roads were constructed both as feeders to the railways or for through traffic across the subcontinent. These means of transport significantly contributed to the development of British India. Initially, roads were supposed to be feeders to the railways. However, in many areas roads were laid out parallel to the railways resulting in competition between the two modes of transport. Moreover, the transfer of goods and raw materials by the railways was affected by the railway freight rates and the tariff rates. Evidently, this promoted the competition further.

Development Infrastructure

A careful study of the data given in Box 8.4 indicates that the British had left behind a huge infrastructure for the development of the two dominions—India and Pakistan-Bangladesh. In this section an attempt has been made to evaluate the base, making a special reference to the railways, roads and other means of over land transportation.

Box 8.4: The British Legacy

1. An integrated territory extending from Peshawar to Burma and from Gilgit and Hunza in the north to Sri Lanka in the south.
2. Skilled and experienced secretarial staff; a civil service to administer the country in every nook and corner of the country.
3. An armed force consisting of the Army, Navy and the Air Force; construction of seaports and the civilian airports, thus ensuring the national security.
4. Development of shipping with acquired ships and other paraphernalia; merchant navy and warehouse.
5. An educational system to train the army, navy and the air force; nodal national academia.
6. Educational institutions from primary schools to the university and the research laboratories distributed in different parts of the country; their patronage of the low-level *pathshalas* and *maktabs*; a network of engineering colleges; training centres, universities at Allahabad, Lahore and Calcutta, presidency colleges.
7. Development of port centres, such as Madras, Calcutta, Bombay and Karachi; which emerged as the throbs of the economy and culture; and the centres with the management of financial institutions and banks.
8. Police, PAC and para-military force; the country was divided into various regional centres for the training for civil defence.
9. An administrative set-up to govern the country; training of the staff from the *tehsildar* to the sub-divisional magistrate and the commissioners; district magistrate and the collector.
10. A judicial system from the *munsif* at the *tehsil* level and the district judge to the high courts at the provincial level; an apex court.
11. Development of knowledge collection organisation, e.g. Survey of India (SOI), Geological Survey of India (GSI), Anthropological Survey, Archaeological Survey, Zoological Survey of India.
12. Development of agriculture, animal husbandry and rural arts; a revenue administration for collecting land tax, settlement reports, cadastral surveys and publications of cadastral maps; administrative service from *patwari* to the sub-divisional level, *tehsil* level and the district level; provision for revenue collection and redressal of land disputes.
13. Development of postal service.
14. Publication of the reports based on the work of the different surveys mentioned above; memoirs and land records.
15. Conservation of monuments of historical buildings and other related work and management of display of artefacts and paintings and sculpture in the museums.
16. Introduction of a system for the assessment and collection of land revenue; different land systems and different modes of tax collection suited to the needs of the regions of the country; tenurial systems; settlement reports, themselves a storehouse of knowledge on land systems and agricultural development.

(*Contd.*)

17. Launching river valley projects, construction of a network of canals to protect the drought prone areas from famines; the network of canals, particularly in Punjab and Sind to transfer river waters to areas of chronic rainfall deficiency; canal colonies over land's hitherto considered as waste, the irrigation system transformed the agricultural scene from deficiency of food grains to surplus and production of long staple cotton to reduce dependence on cotton imports.
18. Institutions for the conservation of plants and animals; zoological and wild life parks and reserves.
19. A network of public libraries in all districts covering the whole country.
20. Establishment of the public hospitals and the health care delivery system; education and training of health personnel; institutions for the training and education of the armed forces; division of the country into several commands for operational purposes.
21. Empowering SOI to undertake the Trigonometrical Survey of the country; publication of topographical sheets on different scales from 1:1 million, 1:4 miles, 1:1mile, 1: half a mile; creation of infrastructure for the compilation of atlases and map-making for specialised use and for public use.
22. Authorising GSI to publish memoirs and records on the work of GSI.
23. Authorisation of the ASI to display the findings in museums and art galleries; establishment of training centres for the ASI staff.
24. Construction of jails at the district head quarters; training and administration of jails.
25. Compilation and publication of Gazetteers on each district as well as at the level of provinces.
26. Infrastructure for the administration of customs, excise duty and tax collection.
27. Authorising ASI to conserve monuments, their repairs and restoration work.
28. Appointment of Royal Commission for enquiries and status reports, such as Royal Commission on agriculture, Royal Commission on industry, Royal Commission on other problems to identify the nature of problems and their solutions, creation of financial institutions such as banks, printing and circulation of paper currency and silver coins.
29. Creation of cooperative societies and rural banks.
30. Agricultural demonstration works and the establishment of government agricultural farms.

Source: Compiled by the author.

Railways

British India had a total route-mileage of 46,922 miles or 75,513 km. This was the position on March 31, 1939. Of this total, 29,764 miles or 47,900 km. (63 %) was state-owned and 17,158 miles or 27,613 km. (about 37 %) was directly managed. Economic historians recognise several periods in the construction and management of the railways. The periodisation scheme is as follows: (a) pre-1900; (b) pre-

World War I; (c) during the World War I; (d) post-World War I; (e) pre-World War II during the economic recession; and (f) during the World War II.

It may be recapitulated that the proposals for the construction of railways in India were invited for the first time in 1844. Railway construction companies incorporated in England responded. In this connection reference may be made to the East Indian Railway Company (EIR) and the Great Indian Peninsular Railway Company (GIPR). It may, however, be noted that the capital for railway construction came from England and not from India. Indians were perhaps hesitant of investing their capital in railway construction. The purpose of constructing the two trunk lines, referred to above, was to connect the hinterland of the two Presidencies with the principal ports. These lines also connected the different Presidencies with one another. Contracts were signed with eight companies during the period 1854 to 1860. Large investment of the English capital helped the projects of railway construction. The main features of the contracts with the guaranteed companies were: (a) pre-grant of land; (b) a guaranteed rate of interest—4.5 to 5 %; (c) utilisation of one-half of the surplus profit earned by the companies; (d) reservation of power of supervision and control by the government; and (e) the government reserved its options to purchase the lines after 25 or 50 years. This practice was known as the old guarantee system. This, however, proved to be a burden on the taxpayer in India. The railway budget suffered a huge deficit amounting to about Rupees 16.7 million. The system was widely criticised for being expensive both in India and in England.

The review of the railway policy and the criticism that it faced resulted in two important changes. These changes enabled the government to claim, unconditionally, one-half of the surplus profit every six months. Capital investment up to two million pounds sterling was transferred to the railways annually. A new cheaper gauge—the meter gauge—was adopted. Work on the construction of North Western Railway (NWR) connecting northern India with Punjab and Sind started. In some sectors conversion from meter to broad gauge was found to be necessary for strategic reasons. However, recurrent famines in the period 1874-79 and the war with Afghanistan upset the progress of work. The Royal Famine Commission of 1880 demanded more railway mileage. It was that some 5,000 additional miles (8047 km.) were needed urgently to combat the problem of famines. Contracts were made with the new guaranteed companies, such as Bengal-Nagpur Railway (BNR), Madras and Southern Maratha Railway Company (MSMR). The government wanted to ensure that the lines were the property of the government. It was also proposed that companies should be involved in the construction of the branch lines. The government also invited Indian princely states to undertake construction of railways in their territories. With the beginning of the twentieth century the railway construction activity increased substantially. It was because the government earned large profits. The railway mileage in the pre-World War period increased from 24,752 in 1900 to 34,656 in 1913-14. This period witnessed substantive increase in railway profit. The increase was largely due to the progress in the field of economic development. This was the period when Punjab and Sind irrigation works were completed. They generated new opportunity for railway development. The First World War (1914-18) led to the disruption of the railway system. The railways were asked to transfer troops and materials on a large scale. Besides, the British were involved in the war in West Asia. The war also disrupted the supply of railway material from England.

'There were scores of bridges with girders unfit to carry train loads...There are many miles of rails, hundreds of engines and thousands of wagons whose rightful date for renewal is long over past' (Acworth Committee Report 1931). In February 1923, the government took over the major trunk lines—GIPR and EIR—with the intention of direct management. This step led to further involvement of the government in railway management. In the next two decades, the Bombay, Baroda and Central India Railway (BB and CIR) and the Assam-Bengal railway (ABR) were also taken over with effect from first January 1942. Some sectors of railways were merged. For example, Rohilkhand and Kumaon Railway (RKR), Bengal and Northwestern railway (NWR) were also taken over and amalgamated under the name of the Oudh and Tirhut Railway (OTR). An important recommendation of the Acworth Committee was the separation of railway finance from the general finance.[3]

British India faced the consequences of the World War II (1939-1945). The Indian railways were put under pressure for increased traffic due to the war. The railways were given the responsibility of transporting several items needed for the war. During this period heavy investments were made on the new construction, improvement in working methods and acquiring more powerful locomotives. Sir Ghuthrie Russell, chief commissioner for railways, commented in his budget speech in 1940-41 that the railways with their existing capacity were in a position to absorb the coastwise traffic in all commodities except coal.

According to some critics, railway rates policy was so constituted that it favoured more the European traders and hindered the development of Indian industries. The general impression was that the rates policy encouraged the export of raw materials and the import of foreign manufactures.[4] The Fiscal Commission, Commission on Agriculture and the Industrial Commission agreed that the rating of internal traffic should be done on the basis of equality. 'With traffic of the same class and over similar distances to and from the ports should be the same'. The Agricultural Commission noted that the railway rates had a bearing on agricultural development. The Commission suggested that a global cooperation between the agricultural departments and the railways was urgently needed. The Agricultural Commission also suggested that the rates policy should be re-examined. The transport of fertiliser, fuel, fodder, milch cattle and other related items affected seriously the process of agricultural development. The Wedgwood Committee noted that the Indian railway rates were illogical and called for drastic simplification. Acworth Committee had recommended the formation of the Rates Advisory Committee. It was suggested that India needed a tribunal on the pattern of Railway Rates Tribunal in Great Britain. Another recommendation was that a Railway Fund should be instituted. 'Railways have increased the ruralisation of the country by contributing to the decline of the indigenous industries. The general feeling was that the country needed further development of railways to suit the requirements'. Comparative statistics show that India was far behind many countries in the development of railways. Europe minus USSR had 1.7 million square miles of area under railways and 190,000 miles of railways. As compared to this the two dominions of India and Pakistan had 1.8 million square miles of area and 40,509 miles of railway length.[5]

Roads

The development of roads in British India was much less than desired. In the first stage the East India Company was not interested in road building activities. It is understandable because the company was a commercial corporation. However, Lord William Bentinck evinced great interest in road building. His view was that enforcement of law and order was the necessity and the roads should be constructed on a comprehensive basis. Due to his interest, the Grand Trunk Road (GTR) was repaired and developed. It may be recapitulated that the GTR was a piece of Sher Shah Suri's imagination. The road was lying in a bad condition in several sectors. Moreover, it was not properly maintained. GTR linked Peshawar with Calcutta. Maintenance of such a highway was necessary because of its strategic location. Unfortunately, roads were given under the charge of the Provincial Military Boards. This restricted the full development of roads. It was necessary to have a national-level authority to construct and maintain trunk routes. Lord Dalhousie took interest in road building and his policy was straightforward. His suggestion was that the country should have several trunk roads. As a result four GTRs were constructed which linked Khyber with Calcutta, Calcutta with Madras, Madras with Bombay and Bombay with Delhi.

In 1943, British India had a total length of 95,054 miles of roads. As the railway construction proceeded, the need for roads became more acute. The roads were supposed to be helping the railways. This led to a demand for bridged and metalled roads (*pucca* roads) at right angles to the railways and giving access to them in all the seasons of the year'.[6] British India was a vast territory, but some regions, such as Rajputana, Sind, southern Punjab, Orissa and Bengal lagged behind. There were several obstacles in the construction of roads, such as low density of population, unbridged sectors, lack of land and lack of road materials. Besides, British India had a very large network of *kuchcha* roads, the total length being 201,414 miles (324,144 km.). These were mainly rural roads. They were not fit for travel during the rainy season. The condition of roads was unsatisfactory, more so because the responsibility was given to the local municipal bodies. There was a crunch of capital resources for the construction of roads. It was natural that these bodies did not commit themselves to the construction of roads. Under these circumstances Kenneth suggested an alternative body—Provincial Highway Department. He recommended that the body should be established in every province of British India. In his understanding, this body should look after and carry out the construction of district roads. As the time passed the need for roads became more acute. Moreover, the emergence of mechanical vehicles—motorcar, omni-bus, and trucks—further provided an impetus to road building. Not much can be said on the significance and the value of the roads in a country where agriculture is the backbone of the economy.[7] The policy of linking the roads to serve railways did not meet much success and the on-going competition between the roads and the railways continued. There is no gainsaying the fact that roads are necessary for transporting the food grains, vegetables, fruits, cooking oil, fertilisers and other requirements of the rural population. India, however, was lagging behind among the countries of the world in progress of road development. The country had twenty miles of roads per hundred square miles. On the other hand, the United States had eighty miles of roads per hundred square miles.

Industries

The backwardness of India in the factory industry was a cumulative result of neglect by the British administration. Indians excel in the manufacture of artisan goods—textiles. In the very first phase, artisan industry (handicrafts, cloth, jewellery, gems, etc.) suffered a set back. In fact, it met its end because of the import of industrial goods from Britain. The East India Company and its shareholders in England wanted the slow death of the Indian artisan industry so that manufactured goods from Britain can find a captive market in India. The situation changed with the constitution of the Imperial Department of Commerce and Industry in 1905. The initiative was taken by Lord Curzon to create an industrial base within the country. On his instance, provincial governments, such as United Provinces of Agra and Oudh and the Madras Presidency took initiatives in the industrial activity. However, within a short period of five years in 1910 Lord Morley, the then Secretary of the State, did not give much support to this idea. He cold-shouldered the proposal for indigenous industrialisation. It was argued that the necessary organisation and equipment for the industry should have been created as a private enterprise. The British approach to India's industrialisation suddenly changed during the First World War (1914-18). The war cut off the import of goods from Britain. This situation was difficult to handle. It pointed towards the need for indigenous industry based on the local resources and labour supply at least in sectors like consumer goods, etc., so that the needs of the local population were met. The war, it appears, highlighted the exceptional opportunities for the Indian industry. The break down of the import trade resulted in a change of policy on industrialisation. It was felt that the goods, which formerly were imported from Britain, should now be manufactured locally. There were many difficulties in the way to self-sufficiency in industrial products. First, obtaining materials and essential machinery was not easy in the war environment. Secondly, there was a scarcity of technical experts who could have helped in the early rise of industry. Thirdly, skilled labour was in shortage. Fourthly, railway wagons were not available. Lastly, there was a scarcity of coal as there were no coking plants. The British appointed an Industrial Commission. The British government at last realised after loosing a lot of time that industrial development of India was a necessary pre-requisite for the self-sufficiency in the industrial sector. In the early phase, priority was given to industries like cotton textile, jute textile, iron and steel, leather goods, engineering industry, chemicals, mineral acids, vegetable oils, paper, glass, cement, cutlery, fertilisers, paints, varnishes and surgical instruments. It was a new beginning after the bitter experience of short supplies during the war period. The British administration in India learnt it the harder way. A series of action plans were launched. The departments of industries were created in all the provinces. The Government of India Act of 1919 paved the way for indigenous industries and defined the jurisdiction of the provinces in so far as the industrialisation was concerned. The war was considered as an industrial boom. But it proved as ephemeral as ever. While it is true that the period 1919-21 marked the beginning of the process of industrialisation, the depression came fast. It seriously affected the industrial projects because of the deflation of currency. The year 1920-21 came with a big slump. As the value of the rupee was reduced, the rising prices of agricultural produce also seriously affected the Indian industry. The end came on the heels of the beginning.

Indian industry suffered another shock in 1937-38 when the world economy moved towards another depression. Agricultural prices registered recovery and the circumstances again became favourable for industry. In the first half of 1937-38 several industries registered an increase in production. This particularly happened in sugar, cotton, steel, coal and jute. The tariff policy of the British played a role in creating better conditions for the development of the industry in the subcontinent. The Second World War helped the progress of the Indian industry.

The political scene changed as the Indian National Congress Party pleaded for comprehensive industrial development. As a result, two associations came into being: an all-India Spinners Association and an all-India Village Industries Association. Emphasis was laid on key industries like machinery, plants and other machine tools, automobiles, electrical plants and accessories, heavy chemicals, power generation and supply. The Congress came out with a demand for a comprehensive policy on national planning. With this view in mind, a National Planning Committee was constituted in December 1938. Nehru was appointed as the committee's first chairman. Besides, 29 sub-committees on various subjects were also constituted. Twenty of them reported their progress to the National Planning Committee. The basic idea was to enhance the standard of living of the people not only in terms of the absolute necessities but also in terms of the amenities for civilised existence. The Congress party expected to achieve these goals within a period of five to ten years.

Indian industry found favourable environment of growth during the Second World War. Overseas orders of the tune of rupees 1600 million were received. The orders came for the goods manufactured in India, which included cotton textiles, jute manufactures, iron and steel, engineering and leather goods. The industry had to cater to the needs of the army. The armed forces needed at least twenty thousand articles. The demand gave a fillip to the indigenous industry in India. During the Second World War, Indian industry produced 90 per cent of total war requirements, sending abroad arms, ammunitions, shells, rifles, sand-bags, saddles, electric cables, road rolling plants, electrical goods, fans and greatcoats. The British government in India constituted a Board of Scientific and Industrial Research in 1940.

Levels of Development

Development and Rural Development

An objective accomplishment analysis of rural development in India can possibly be attempted within the framework of dialectical logic. In appearance and design the development policy is supposed to reduce the social disparity; in essence it resulted in the very resurgence of these disparities. The explanation of this phenomenon lies in the internal contradictions of the social order. The hierarchised society has not changed over time. From the beginning of the planning process in India there were two major objectives of development planning: (a) steady economic growth; and (b) equitable distribution of the gains of development. The development policy of the ruling Congress Party as stated by Nehru was: means of production should be socially owned and controlled for the benefit of society as a whole. This was a very socialistic view seen in the context of predominantly agrarian economy and the distortions by the inequitable distribution of land. Although much diluted by exigencies of the changing political situation, the concern for social justice remained at the back of all planning exercises aimed at the development of the rural areas of the country. However, on the completion of six decades of planned efforts, the cumulative impact of the national policies and the series of rural development programmes reveals itself in the two mutually contradictory facets: (a) accumulation (appropriation) of the developmental gains by the privileged few; and (b) the destitution of the many—the downtrodden, underprivileged and the marginalised sections of the society.

Nehru's planning ideals and objectives were in sharp contrast to the political situation in the country. There were several economic constraints, which did not allow smooth industrial development. The international economic order, for example, did not favour comprehensive planning. His idea of achieving the goals of growth with social justice through rural development planning proved ineffective. The society was sharply divided on caste-class lines. The needs of the different and disparate segments of rural population were supposed to be fulfilled without mutual conflict between the classes on the one hand, and the goals of sectoral and regional growth on the other. Some of these constraints were insurmountable: an enormous population with an alarmingly high growth rate, socio-cultural and ethno-lingual diversity, inequalities in the distribution of land, disparities in the distribution of natural resources and the degree of their exploitation, highly varied levels of technology, gaps in the diffusion of new technology and the associated modernising process, persistence of the dual economy with the conflicting objectives of the private and public sectors, and the subservience of the long range national objectives to the immediate political interests of parties and individuals. In retrospect, one cannot escape the conclusion that the nation did not have the political will to cope with such a complex situation. While rural development programmes continued to be a major policy plank in states and central governments, the accent on an 'integrated approach' is a recent phenomenon which developed as a result of a United Nations General Body Resolution in 1970. It is necessary to evaluate the social implications of national planning policies on rural development, placing it in the perspective of the stark realities of the rural social and economic order, and unravelling the underlying ideological premises and their internal contradictions (Ahmad 1987).

Demographic Indicators

The subcontinent extends over an area of 5.22 million sq. km. The total population of the countries in South Asia as counted in 2000 A.C. was 1,367.18 million (Appendix V). Out of this, India alone accounts for 63.03 per cent of the total area and 74 per cent of the total population. On the other end of the scale is Maldives. Its share in area is 0.01 per cent and in total population 0.02 per cent. After India, Pakistan is the second largest country and the second most populous country among the South Asian nations. In terms of area Pakistan's share is 16.87 per cent of the total and 11.45 per cent of the population. Looking at these statistics one cannot escape the idea that India is like a giant in the company of her little neighbours. Inevitably, Indian governments have always displayed a massive sense of superiority. This is the main reason why all other countries of the South Asian region are apprehensive of the risks involved (Table 8.1).

Table 8.1: Area and Population of the Countries of the Subcontinent

Sl. No.	Country	Population* (in '000)	Proportion to Total Population of the Subcontinent (%)	Area (in sq. km.)	Proportion to Total Area of the Subcontinent (%)
1.	Afghanistan	22,720	1.66	645,806	12.38
2.	Pakistan	156,483	11.45	880,254	16.87
3.	Nepal	23,931	1.75	147,181	2.82
4.	Bangladesh	129,155	9.45	142,615	2.73
5.	Bhutan	2,124	0.15	46,673	0.89
6.	Sri Lanka	18,827	1.38	65,610	1.26
7.	Maldives	286	0.02	298	0.01
8.	India	1,013,662	74.14	3,287,263	63.03
	Total	1,367,188	100.00	5,215,700	100.00

Source: * WHO 2000.

Population Growth Rate

The demographic history of the subcontinent indicates high growth rates of population. It is true in the case of Afghanistan where the population is growing at the rate of 3.0 per cent per annum. Growth rates of other countries are as follows: Pakistan 2.9 per cent, and Nepal 2.5 per cent. The population of other countries in the region, such as Bangladesh and Sri Lanka has been growing at the rate of 1.7 per cent and 1.1 per cent respectively. It shows that these countries have taken steps to curtail the growth rates of population fairly successfully. On an average, India's population has been growing at the rate of 1.9 per cent annually in the recent decade. Other countries, such as Maldives and Bhutan experienced a higher population growth rate of 2.9 per cent and 2.4 per cent respectively. However, the net addition of population in these countries was too small to reckon with. The countries can be arranged in a descending order according to the population growth rates.

Table 8.2: Ranking of Countries on Demographic Indicators, 2000

Rank	PGR	UP	GRUP	GRRP	DP	LEB	
						Male	Female
1.	Afghanistan	Pakistan	Bhutan	Afghanistan	Bangladesh	Sri Lanka	Sri Lanka
2.	Pakistan	India	Afghanistan	Bhutan	Maldives	Maldives	Pakistan
3.	Maldives	Maldives	Nepal	Maldives	India	Pakistan	Maldives
4.	Nepal	Bangladesh	Pakistan	Nepal	Sri Lanka	India	India
5.	Bhutan	Sri Lanka	Maldives	Pakistan	Pakistan	Bhutan	Bhutan
6.	India	Afghanistan	Bangladesh	Bangladesh	Nepal	Bangladesh	Bangladesh
7.	Bangladesh	Nepal	Sri Lanka	India	Bhutan	Nepal	Nepal
8.	Sri Lanka	Bhutan	India	Sri Lanka	Afghanistan	Afghanistan	Afghanistan

Notes: PGR: Population growth rate; UP: Urban Population; GRUP: Growth rate of urban population; GRRP: Growth rate of rural population; DP: Density of Population; LEB: Life expectancy at birth.

Source: Table compiled by the author.

Urban Population

In so far as the level of urbanisation is concerned, there are significant differences among these countries. Pakistan is the most urbanised: 37 per cent of population lives in the towns and cities. About 25 per cent of population of Sri Lanka lives in towns/cities. The same is true for Bangladesh and the Maldives. Both Bhutan and Nepal are least urbanised with the respective figures being 7.1 per cent in Bhutan, and 11.9 per cent in Nepal. India, with an urbanisation level of 28.4 per cent is far behind Pakistan. The countries can be arranged in a descending order according to the percentage of urban population (Table 8.2).

Growth Rate of Urban Population

In the case of India the base of population is so large that even the slow rate of growth of 2.7 per cent is likely to add to a large quantum of urban population. Annual growth of population in other countries of the region is as follows—Bhutan 5.8 per cent, Afghanistan 5.3 per cent, Nepal 4.9 per cent, Pakistan 3.9 per cent, Sri Lanka 3.1 per cent, Maldives 3.8 per cent and Bangladesh 3.7 per cent. These data show that the process of urbanisation in these countries is quite fast. People have been flocking to the urban centres because the rural areas have very little to offer. These countries are progressing fast in so far as the rate of urbanisation is concerned; land-locked states of Afghanistan, Nepal and Bhutan are in the fast forward lane. But the size of population in these neighbouring countries is small and the growth rate of urban population is not expected to bring a dramatic change in the near future (Table 8.2).

Growth Rate of Rural Population

The population in these countries is mainly concentrated in the rural areas. The masses are dependent mainly on agriculture and allied activities. The growth of population in the rural areas is likely to grow substantially. The growth rates of rural population in these countries are low. However, Afghanistan (2.5 per cent), Bhutan (2.2 per cent), and the Maldives (2.1 per cent) have registered high growth rates. But the net addition to rural population is not going to be significant because of the low population base. Moreover, these rates have been projected for the period 2000-2015. During the same period India registered a growth rate of 0.5 per cent, Bangladesh 0.6 per cent, Sri Lanka 0.2 per cent, Nepal and Pakistan are likely to register higher growth rates—Nepal 1.6 per cent and Pakistan 1.2 per cent (Table 8.2).

Density of Population

Another dimension of the question of growth of rural population may be explored by citing the population density on agricultural land as expressed in population per hectare. This is also described as physiological density. These figures are alarming. The densities vary from 9,922 for Bangladesh, 9,533 for Maldives, 3,409 for India, 2,913 for Sri Lanka, 2,030 for Pakistan, 1,673 for Nepal, 452 for Bhutan to 348 for Afghanistan. It may be noted that the figures for Maldives may be misleading because the population is largely dependent on marine resources, agriculture being a subsidiary activity. It may, however, be noted that the hill economies, such as those of Afghanistan, Nepal and Bhutan have alternative occupations, e.g., artisan industry, pastoral grazing and activities related to tourism (Table 8.2).

Life Expectancy

Life expectancy at birth is computed by an application of analytical methods. The data may be derived from statistical analysis. It is in this respect that the statistics cannot be seen on their face value. The data are not on sure footing because of the problems of authenticity. The data were computed for the period 1995-2000. The data reveal sharp differences among the South Asian countries. Let us analyse the data on male life expectancy at birth. The figures vary from 45 years in Afghanistan to 71 years in Sri Lanka. On an average the males in Sri Lanka can think of living up to 71 years. The comparable figures for Maldives are 66 years, 63 years for Pakistan, 62 years for India, 60 years for Bhutan, 58 years for Nepal and Bangladesh. Afghanistan with 45 years of life expectancy is at the bottom.

In so far as longevity is concerned females in the South Asian countries have an edge over the males. The data show that the female life expectancy at birth is more than their male counterparts. However, the male-female gap in Nepal is marginal. In all other countries women may expect to live more than males. The data show that the female life expectancy is the lowest (46 years) in Afghanistan, 57 years in Nepal, 58 years in Bangladesh, 62 years in Bhutan, 63 years in India and the Maldives, 66 years in Pakistan and 75 years in Sri Lanka.

It can be concluded that Sri Lanka has been doing well in regard to life expectancy at birth and is at the top. The respective regimes in Sri Lanka have gone ahead for promoting longevity through effective health and nutrition programmes. Afghanistan is at the bottom. Another feature is the higher longevity of the female population. There are striking differences in male/female longevity in Sri Lanka, Pakistan and Bhutan. On the other hand, no significant differences have been observed in Afghanistan, Bangladesh, India and Nepal.

Human Development Indicators

The quality and the quantity of safe water supply are considered as one of the most critical indicators of human development. In certain parts of India, like Rajasthan, women have to go to long distances to collect drinking water. Sometimes the distance may be 10 km. It is the responsibility of the women to go out to fetch water. An equally important indicator is access to improved sanitation. We have tried to analyse here the data on crucial quality-of-life indicators, like access to safe water and sanitation.

Access to Improved Water Supply

Maldives stands first with its total population having access to safe water supply. The situation is worst in Afghanistan where less than 15 per cent of the population has this amenity. India, Nepal, Pakistan and Bangladesh are better off in terms of the supply of safe water (Table 8.3).

Table 8.3: Access to Improved Water Supply (%), 2000

Population	*Rural*	*Total*
Maldives	100	100
Bangladesh	97	97
India	86	88
Pakistan	84	88
Sri Lanka	80	83
Nepal	80	81
Bhutan	60	62
Afghanistan	11	13

Source: WHO 2000.

Access to Improved Sanitation

In terms of access to improved sanitation the position in India is as bad as in Afghanistan. In the same way Sri Lanka is on the better side of the human development scale. Bhutan, Pakistan, Maldives and Bangladesh seem to be in better position as compared to India (Table 8.4).

Table 8.4: Access to Improved Sanitation (%), 2000

Population	*Rural*	*Total*
Sri Lanka	80	83
Bhutan	70	69
Pakistan	42	61
Maldives	41	56
Bangladesh	44	53
Nepal	20	27
Afghanistan	8	12
India	14	—

Source: WHO 2000.

Literacy and Education

The educational development of the people has a history of 200 years when the British imposed a system of education in the Indian empire. In fact, indigenous system of education before the British was confined to *pathshalas* and *madrasas*. The two institutions are continuing as such. Their course content and coverage was found to be inadequate. We have examples of old institutions, such as *madrasa* Safdarjung and the *madrasa* in Hauz Khas. There was an extensive expansion of indigenous education. One may refer to the Hauz Khas *madrasa*, which was catering to the needs of the education of both Indians and foreigners. There were students drawn from all over the continent. They were also serving as residential institutions. The development of education under the *Raj* had several regional dimensions. As the history of the British expansion in India shows, the first people to be exposed to modern education were located on the periphery of the subcontinent. For example, presidency colleges were established in Calcutta, Madras and Bombay. This dimension of development hurt the traditional educational system. Regional disparities emerged out as phenomena. This was partly due to the fact that the periphery development and the core areas, which were historically important, started decaying. The first people to be exposed to modern education by the British were the Bengalis and the Madrasis. They joined the administrative service in different sectors of governance, while the people in the core areas were lagging behind.

The development of modern system of education has left much to be desired. About one-fourth of the population of the subcontinent in the age cohort of 15 years and above is literate. This is the major indicator of educational backwardness. It is a silent reminder that immediate action is needed for human development. Let us review the position of education in the states of the south Asian subcontinent.

While primary education is free for all in the age cohort of 6-10 years in Nepal, not all children are registered at the primary school level. There are constraints, both social and economic, which do not make it possible for the children of the poor and the underprivileged families to avail this facility. The total enrolment at the secondary level is 0.8 million. Tribhuvan University in Kathmandu caters to the needs of the people at the higher education and research level. It holds no promises for the poor and the deprived sections of population. The Tribhuvan University has several campuses.

In Pakistan the overall literacy rates are not very encouraging. Only 45 per cent adults are literates (males 59 per cent; females 30 per cent). The school enrolment data show a backlog to be cleared by the state. An idea can be formed from the fact that schooling at different levels reflects disparities. Recent data show that only 8.9 million children are enrolled at the primary level and 3.5 million at the secondary level. This shows a massive dropout from primary to secondary levels. An estimated dropout rate is no less than 5.4 million. These statistics throw light on the state of the art. The universities located at Karachi, Lahore, Peshawar, Dadu, and Faisalabad together have some 6.68 million students on the rolls. Pakistan has an open door policy in higher education and research. If better education is available elsewhere, why not avail it!

A recent trend is that the best educational institutions in the western world are establishing their branches in the South Asian countries. The universities in the West have found this opportunity most lucrative. Take for example the case of India, where Canadian, American, Australian, Chinese and the British institutions are establishing their branches to attract the student population. The idea is that they should receive quality education in their home country rather than going abroad. There is no need for those seeking higher education in the West to go abroad. This is due to the fact that these institutions are there at their doorstep.

India today is passing through a crisis. Stresses and strains generated by several sections of the society, on the one hand, and regions (read states) on the other, have never been so intense. The source of disenchantment is common: inability to live a better life. This is possible only if the developmental strategies are so designed that they shed their urban and elite bias. Education is being looked upon as a great hope. This realisation has come too late. The educational policy and the preferences are now focusing on new possibilities. Education is no more considered as a 'social service' alone but as a significant ingredient of development. It is treated as a commodity. This, however, seems to be a challenge in the contemporary situation wherein over-emphasis on formal education directed towards passing of examinations tends to undermine the scope of education itself. It was unfortunate that the role of education in improving the quality of life lost sight at all stages. The advancement of science and technology and their role in development, while creating social tensions and conflicts, has put additional responsibilities on the sector of education as managed by the states. The challenges of social distance, alienation and deprivation of the masses are now being resolved in a development frame.[8]

These are aspirations to be translated into reality. In fact, these aspirations were to be fulfilled through the process of developmental planning initiated three decades ago. However, hopes were belied because the planning process revealed its class character very soon and those who were deprived were further pushed into the realm of underdevelopment.

Illiteracy Rate

Other indicators, such as illiteracy rates, hospital beds per mille and population below the poverty line have been described as crucial in any evaluation of the state of human development. The term illiteracy is a negative indicator. Instead of examining the data on literacy, illiteracy has been given preference over literacy. One may argue that data on illiteracy show the quantum of work to be done to eradicate the rampant illiteracy. The percentage of population aged 15 years and above has been covered by the literacy drives launched by the national governments in each country. The assumption is that population cohort below 15 years of age is by and large exposed to schooling. In Afghanistan 64 per cent of the population is illiterate, the male-female break-up being 49 and 79 per cent respectively (Table 8.5). In Bangladesh the overall illiteracy rate is 59 per cent. The male-female break-up is 48 and 71 per cent respectively. In Bhutan the aggregative position is that 53 per cent of the population lives below the literacy line (39 per cent males and 66 per cent females). As compared to this in India the overall illiterate population is 44 per cent (31 per cent males and 58 per cent females). In the case of Nepal, overall illiteracy rate is 59 per cent (41 per cent males and 76 per cent females). In Maldives overall illiteracy rate is 4 per cent (4 per cent each for males and females).

Table 8.5: Illiteracy Rates (%), 2000

Population	*Total*	*Male*	*Female*
Afghanistan	63.7	49.0	79.2
Bangladesh	59.2	48.3	70.5
Bhutan	52.7	38.9	66.4
India	44.2	31.4	57.9
Nepal	58.6	40.9	76.2
Pakistan	56.7	42.4	72.2
Sri Lanka	8.4	5.5	11.1
Maldives	3.7	3.7	3.6

Source: Mahbub-ul-Haq, 2002.

Impact of Literacy on the Marginalised Tribal Communities

The tribal population of India may be selected as an example of neglect in terms of social and human development. This segment of population has suffered in the wake of planned economic development. The negative impact of development is the result of mismatch between social and economic formations of the tribal societies and the planning strategies. A healthy process of growth can be initiated if the development process is endogenous and grows organically within and is not thrust from without. The exogenetic forces of change do not produce desirable results. In fact, they result in the alienation of the tribes and if not manipulated properly are bound to destabilise the tribal social and economic order.

Education can play a catalytic role in checking some of these distortions of planning, but the way educational programmes were implemented in tribal areas not only proved irrelevant but also made education an easy distraction. It was considered irrelevant because it could not help them in improving their life situations, it became an easy distraction because it came in the way of earnings through family labour. The fact is that the literacy rate of the tribal females in urban areas is only a broad generalisation. The regional expression of the on-going process even questions the existing developmental efforts. The literacy rate of 0.93 per cent in case of rural females in Rajasthan should be enough to prove the failures of the educational planners. Even in the formal system of education, the success has been very little as is evident from retention rates at primary stage. Evidently, the tribal population is not able to cope with the rate of change in other communities, not even with the scheduled caste. The reason for this stagnation lies in the tribal habitat characterised by isolation—physical as well as social—which does not allow the forces of change to make a breakthrough.

Seen in this background, the transformation of tribal societies contrasts well with other segments of population in all spheres of development, particularly education. The tribal literacy in 1981 compares well with general literacy in 1941. This time lag may not look significant in the context of the process of social evolution, but it has serious implications for national efforts to build an egalitarian society.

A point of caution is necessary here. Low level of literacy or for that matter low educability should not be confused with their capacity to come to terms with their environment for generating basic minimum for their existence. Even the basic minimum should be viewed in the context of tribal formation because it is far from minimum in the context of development. This, however, has acted as a constraint on acquiring new means of production in order to make better use of national resource endowments. Education, if planned properly, can play a catalytic role in this interaction. The planning should flow from the context of their immediate habitat on the one hand and from roots of deprivation on the other. Viewed in this context, the problems of educability of tribes are complex. The nature of the problem is so different that it cannot be compared either with the educability of the general population or with the scheduled caste segment. It may be hypothesised that education is an instrument of social change. Its very acceptability is caused by the state of the society. By and large, the spread of education among the tribal communities in India, Pakistan and Bangladesh depends, among other things, on their capacity to receive education, their stage of social evolution, habitat or ecological setting, mode of economy and their exposure to a language other than their own dialect. The process of economic development and the pattern of interaction with the non-tribal groups in the neighbourhood of the areas of tribal concentration may lead to the socio-cultural transformation of the tribes. In this connection the role of the tribal elites, if any, cannot be ignored, as their attitudes tend to operate as a constraint on the universal distribution of the gains of educational development.

The tribes have been exposed to literacy only recently. By and large, their response to programmes of literacy and of formal education has varied significantly between the tribes and among the regions in which the tribes are concentrated. These responses depend on their socio-cultural, economic and demographic characteristics and on the magnitude and direction of the processes such as urbanisation and industrialisation. The influence of Christianity in some tribal areas has also played a significant role in their upliftment. The initiatives taken by the government, other quasi-government and voluntary organisations with the objective of educating tribes through programmes of special education and literacy drives have also contributed significantly towards the process of social change. However, these forces have not operated uniformly with equal vigour in all the tribal areas, thus giving rise to regional disparities.

Population Below Poverty Line

The phrase poverty line is often used by the writers on quality of life and human development. However, estimation of poverty is possible only with the help of statistical techniques. All criteria applied in the measurement of poverty are related to wages as well as the prices. Even after statistical analysis the data are not free from faults. Several writers on this theme point out that the concept of poverty is based on assumptions which are likely to be subjective. Some believe that wages earned is one dimension of reality; the other is the daily diet and caloric intake computed through data which are very often not available. Here an attempt has been made to display the extent of poverty as defined by the data collecting agencies.

Estimates computed for the period 1987-98 reveal the incidence of poverty. About 35-36 per cent population lives below the poverty line in India and Bangladesh. The respective figures for Pakistan are 34 per cent, Sri Lanka 35 per cent and Nepal 42 per cent. The South Asian region, particularly India and Bangladesh have evidence to show that poverty is a common phenomenon. It can be controlled if states in this region collect data not only on income but in consumption level also. The situation is gloomy keeping in view the fact that more and more population is sliding down the poverty line. This is a critical indicator and much depends on how to compute poverty. It is true that the processes of globalisation and the so-called economic reforms are affecting adversely the masses in these countries. General estimates indicate that more than 40 per cent of the population is living in a condition of acute poverty and these figures are likely to increase in the future. Are governments partly responsible for these failures in the eradication of poverty or not? This can also be examined by expenditure on health-related programmes as percentage to gross domestic product (GDP). The figures are not available for India and Afghanistan. The respective figures for the other South Asian nations are as follows: Sri Lanka 2.6 per cent, Bangladesh 3.5 per cent, Pakistan 3.9 per cent, Nepal 5.5 per cent, Bhutan 7 per cent, and Maldives 11 per cent.

Economic Development Indicators

The social roots of the on-going political crisis lie in the under-developed economy and the politics based on it. Primarily, the reason for turmoil is that these problems have not been addressed properly. There are awning gaps in the social distribution of land. As a result the skewed land distribution feeds the engine of revolt. Maoists in Nepal thrive on the lack of official initiatives to remove underdevelopment, particularly unbalanced distribution of land. Like the other developing countries of Asia and Africa, an overwhelming proportion (80 per cent or more) of labour force is clinging to agriculture. Gross national product (GNP) is estimated at US $ 3.3 billion or US $ 170 per capita. Like other countries of the South Asian region, Nepal's fiscal deficit is large. It is evident from the fact that while the domestic revenue is of the tune of US $ 457 million, the expenditure is estimated at $ 725 million. This leaves a gap of $ 268 million. The international trade under the present circumstances is unbalanced. The kingdom exports goods and natural produce to the extent of US $ 376 million; imports have been of the tune of $ 752 million. Since the Himalayan kingdom is a land-locked state, all goods move through the Indian Territory. This is understandable as the network of roads in northern India makes it easy to reach Nepal.

Economic growth rate in Pakistan has moved up from 5.1 per cent during 1965-1980 to 6.1 per cent during 1980-1992. However, poverty has not been successfully brought under control, nor the rigour of the economic hardship of the people has been successfully dealt with. The political leadership usually blames the fast growth of population and a series of *coup d'états* and a prolonged secession war in the Eastern Wing also contributed to the economic instability. Economic surveys have estimated that the GNP had reached the peak mark of US $ 50 billion in 1990, the GNP per capita remaining at US $ 420. Agriculture contributes about 30 per cent of the GDP; the share of the manufacturing sector is around 20 per cent. Broad estimates indicate a gap in the state earnings and expenditure—revenue stands at US $ 9.4 billion, and expenditure at $ 10.9 billion leaving a ghastly gap of $ 1.5 billion. Pakistan has also registered progress in external trade—exports amounting to US $ 6.9 billion and imports amounting to $ 9.6 billion. Evidently, the nation will have to make an extraordinary effort to balance the export-import gap.

A Resume

The concept of development has been borrowed from the life sciences. Organisms like plants and animals, including man develop from a foetus to the newly born babies and the subsequent growth after they are born. All one-day babies of the human kind grow after they have been born.. The development of the body and the mind is sometimes slow and sometimes fast. The development depends upon nutrition, income levels of the family, and heredity. All one-day babies are similar. But they develop significant changes in their shape and body. A one-year old child is differentiable from others on the basis of body size and facial features, skin colour, and hair formation. Used in this sense, the word development implies an irreversible change. One cannot reverse this process because a one-month old child cannot be reverted to a one-day old baby. The process of development in human society is not necessarily irreversible. Historical evidence is there to show that the human society moves forward as well as backwards. The Indian tribals who were exposed to a primitive form of agriculture can be pushed in the reverse direction. If they are ousted from the river valleys, they can return to the primitive stage of hunting and collection from the nature in the surrounding hills and plateau surfaces.

An effort has been made to define development in all its dimensions, such as area development, regional development, economic development, sectoral development, etc. Issues have been raised like sustainable development and human development. Human society once developed, leads to extraordinary use of natural resources with the result that the eco-system is disturbed. The paths of development can vary from the capitalistic models to socialist approach as exemplified by the erstwhile Soviet Union. There is a difference between the state undertakings and the private enterprise. This model has been widely used in the contemporary world. Development of underdevelopment has been largely blamed for biases where capitalist and developed countries of the world exploit resources either through trade or colonial model in which resources of the developing countries are transferred to the developed countries. This is a second alternative and is often described as suction mechanism.

Contemporary economic and regional development conforms admirably to the capitalist model. China has its own model of socialist development. As compared to this, the Indian subcontinent presents a model of petty capitalist system. The nations of the subcontinent have shifted their positions from socialism to capitalist development and *vice versa*. A review of the history of the development of the subcontinent reveals anomalies and disparities in development. The terminology itself is vague. Sometimes terms such as tribal development, hilly area development, women's development, valley development, rural development and urban development are misleading. Development is, or should be, comprehensive and cannot be divided into separate categories.

It is noteworthy that the colonial model as applied by the British in the subcontinent was not always destructive. The British left behind a legacy of development infrastructure, sufficient enough to provide a stable base of economy. Its further development was critically based on this edifice.

The latest focus on human development is as vague as development itself. We have tried to assess the process of human development in the light of some critical indicators, such as access to safe water, sanitation, health, literacy and education.

END NOTES AND REFERENCES

1. According to the official estimates, some 260 million people of India live below the poverty line. Non-official estimates point out a larger figure, say from 300 to 400 million.
2. World Commission on Environment and Development 1987, Our Common Future.
3. 'The dependence of the railways on the exigencies of the general budget prevents them from being run on a commercial basis. This arrangement which assumed that the railway concern goes out of business on 31 March every year and begins life afresh at the beginning of each official year was obviously detrimental to the railways.'
4. 'This complaint was voiced by Sir Ibrahim Rahimtulla in the Imperial Legislative Council in 1915', *cit.* Jathar and Beri 1949: 138.
5. Data for 1944-45.
6. Road Development Committee Report, *cit.* Jathar and Beri 1949: 143.
7. Royal Commission on Agriculture, 1928.
8. Aijazuddin Ahmad, S.C. Nuna (1986): 'Tribal Education: Shadows of an Uncertain Future', in S.C. Nuna, (ed.) *Journal of Educational Planning and Administration*, NIEPA, Vol. I, No. 2, April 1987, pp. 90-107, South Asian Publishers Pvt. Ltd., New Delhi.

9

Epilogue: Reflections on Political History, Development and Poverty

A Summing Up

The people of the subcontinent have multiple identities. However, this cannot be described as pluralism. We cannot say that ours is a 'pluralistic society'. Pluralism may be characteristic feature of African societies, like Kenya. The Kenyan society today consists of three segments: the indigenous tribes, the people of the Indian origin and the people of the British origin. These elements of population present a case of pluralism. Pluralism is an additive concept. This is simply because they do not have common cultural roots. Unlike Kenya, the South Asian subcontinent shows an amalgam of cultural elements, which have evolved during the millennia of inter-mixing and integration. The people have a common heritage of culture, language, dialect and a synthesis of higher cultural elements. Ethnicity, caste-based hierarchy and linguistic multiplicity have all merged together, so much so that it is impossible tò discern them as separate entities. For example, tribes have inherited language and material cultural tradition, which is a product of synthesis. Any amount of analytical methods can disturb sharply distinguished desirable results. They live in enclaves with a certain degree of isolation. This has happened because of the inter-mixing over centuries of living together. It would be difficult to separate the Muslims or the Christians from the Hindus. Similarly, a citizen of the subcontinent may be a Hindu or a Muslim but sometimes he is a Bengali or a Gujarati; he may be a tribal as well as a Hindu; he may be a scheduled caste or a Brahmin of a high rank in social hierarchy. At the same time they have a common history. For example, the cultural tradition of Kerala has been imbibed by all Malyalis irrespective of the fact that they are Muslims or Hindus. There is a cultural regionalism which has made them what they are. We cannot look at them differentially because the Malyali tradition has moulded them into a single whole. They speak the same language. They have a common economy and a common heritage of culture and tradition. One can posit that their history has made them what they are. Evidently, it would be too difficult to define and compare the Muslims and the Christians of Kerala with the Muslims and the Hindus of Uttar Pradesh or Delhi. They are worlds apart. Moreover, our view is that an all-India Muslim identity will be a gross distortion of the social reality. Much has already been said about the historical antecedents of the people who belong to different communities. What binds them together is the common cultural and linguistic heritage and religion is not a basis of distinction and categorisation.

The people of the subcontinent live in their own culture-worlds. Citing the example of Kerala one can say that changes in religion or changes in names and the ways of worship cannot divide the people. The Malyali cultural heritage is shared by all religious groups. The idea that people have inherited different traditions of society and polity would be a gross mistake. The purpose of democracy is to strengthen the individuality of the individual, not the rejection of the specific identity. It is basic and fundamental. Let us pursue the Kerala example further. The Nairs of Kerala, the Moplas and the Christians are basically the same. They are first Malyalis and then anything else. These remarks are equally applicable to all other culture-groups of the subcontinent. The elements of identity are the same in other cultural regions, such as Punjab, Bengal, Orissa or Kashmir. In our view it is not the unity superimposed from above but the diversity which has sustained the states of the South Asian subcontinent. They are unified despite the diversity in culture.

As we know the people of the subcontinent acquired their ethnic and racial characteristics over a long period of time. Geographical factors have played a crucial role in shaping their history. The natural environment consisting of surface structure, relief, and climate have operated as determining factors. In fact, these factors have operated on a combined basis. Geography has shaped their way of life, their economy and regional consciousness. Perhaps, geographical accessibility is the basic principle in regional differentiation. Regions have emerged due to the pattern of spatial interaction and the degree of isolation. The river valleys from Indus to Brahmaputra and from Ganga to Kaveri have acquired their specific traits during the process of peopling of the subcontinent. These river valleys have emerged as acumen of population, agriculture and the cultures based on this. The river valleys have provided space for the settlement of the primordial agrarian communities. It is noteworthy that several parts of the subcontinent have remained by and large cut-off from the rest of the Indian world. The river valleys, on the other hand, have been exposed to both the external and the internal influences. Different streams of migration originating from central Asia and west Asia brought diverse groups in terms of race, ethnicity and culture. During the different phases in our history the people originating in different parts of the world, such as Central and West Asia, Mongolia and the east African plateau, have reached the subcontinent. All this evidence shows that the cultural diversity is an end-product of centuries of living together. Subbarao's view is that the river valleys were colonised time and again by the people who were familiar with agriculture and remained in contact with the successive waves of migrating groups who came after them. There is an interactive model in geographical accessibility and the natural-resource base.

On the periphery of the river valleys, hills and the plateau surfaces have been colonised temporarily by nomadic groups who were always on the move. An example may be cited here of the case of Punjab which has served as a corridor for the migrating groups. Those who came in the earlier wave settled in the river valleys of the north and the west. They were, however, replaced by the subsequent waves. This process of colonisation and settlement has remained a phenomenon throughout our history. The people settled in the land on the banks of the rivers and left imprints of their culture and practices. When a second wave came they were displaced and their place was taken over by these groups. The history, therefore, suggests an ever-lasting phenomenon of settling and displacement alternatively. For example, the river valleys of the Punjab acted as a region of attraction. But the surrounding regions, the highlands and the series of hills in the north and west were not basically suited to the mode of agricultural production. These regions were visited by the nomads who came along with their animal stock and went away. To the south of the Punjab river valleys was located an arid zone which did not attract them. Historians like Toynbee have suggested that older cultures survived in the hills and desert land on the periphery of the regions with a potential of agriculture. Both ecology and race have played a crucial role in ethnic and cultural overlap and differentiation. It is believed that the baffling diversity in population composition, ethnic elements, physical appearance of the people, their life styles, belief systems and the other cultural attributes have largely been influenced by the natural environment.

Any reconstruction of the historical geographies of the parts of the subcontinent shows the process of evolution of the space and its territorial organisation. The subcontinent has passed through a long chequered history of peopling, chieftainships, rise of kingdoms, and empires. *Janapadas* have emerged and died. Kingdoms came and disappeared. Vast empires emerged, expanded in area and then declined. The *Janapadas* have always remained perennial. We have tried to select critical points of time to understand the rise and fall of empires. On the other hand, we have selected the Mauryan Empire of the third century B.C.; Harsha Vardhana's Empire during the first half of the seventh century; Tughlak's Empire in the early fourteenth century; and the Vijayanagar Empire in the sixteenth century A.C. There is a law of the land, which operates as a corrective mechanism. Their rise, expansion and decline have been governed by the same law. Every empire was an example of political super structure but its rise has not affected the lower-level political structure, like the *Janapadas*. The king emperors were temporarily powerful. They played a hegemonic role. The kings and emperors were holding a suzerain authority while the local kings were autonomous and their sovereignty remained unaffected. It is this ding-dong battle between the kingdoms and the empires that characterises the political history of the subcontinent. We have tried to analyse the evidence on how the foreign visions of India were formed. One observes a definite conformity between the foreign and the local notions of space and territory of the subcontinent. Unification and dismemberment have moved side by side but the history has been mysterious. Change and continuity have progressed hand in hand. As the political history shows the art of ruling and governance is a kind of balancing act belonging to the centrifugal and centripetal forces. Today no one can say that the political map of South Asia is complete in its basic function and regional structure. The map is bound to change with the changing time.

The racial/ethnic composition of the present-day population of the subcontinent shows that the different social elements have been derived from various ethno-lingual strands who came at different points of time. The initial settlement of the primeval groups shows how the space factors operated behind their preferences for different types of ecologies. A perusal of the information on the racial stock leads to the conclusion that the early streams of migrants were pushed eastward and southward. This eastward and southward thrust was generated by each subsequent wave of the wandering people. Whatever information is available on the racial stocks have to support the view that the older stocks were squeezed out in the mid-Indian region as well as in the upper valleys bordering the rim of hills and highlands. The same argument can be further developed to show that those who arrived towards the end of the peopling history are concentrated close to the entry points in the Hindu-Koh mountains, sub-Himalayan ranges and their offshoots in Sulaiman and Kirthar ranges and Baluchistan with the result no conclusion can be drawn. The weft and the web of social fabric shows that the earlier streams to reach the subcontinent were called the Austroloids along with the Mediterraneans. In fact, the Mediterranean stock came in several waves: a Paleo-Mediterranean brand, the mainstream of the Mediterraneans and the Oriental group. Their proximity to the bordering region shows that they are among the later streams of migration. As one moves eastwards on the North Indian Plain one discovers a residue of the earlier streams. The same phenomenon seems to have operated in the movement of people towards the south, particularly the region south of the Narmada and the Mahanadi rivers. The Indian tribes who account for about 8 per cent of the total population consist of several racial strands, like Proto-Austroloid stock mixed with several strands of the Mediterraneanoids and the Austro-Asiatic elements. The Indian tribes are heavily concentrated in the hilly regions of the mid-India from Sahyadris, Vindhyachal, Satpura, Kaimur range and the Chotanagpur plateau, the newly created state of Chhattisgarh and the Jharkhand. The tribes are also concentrated over the area from the Orissa Highlands to the Rajmahal hills. In the northern and the eastern parts of the subcontinent in the sub-Himalayan region lie a Mongoloid stock represented by several tribal groups like the Laddakhis, Kinnauris, Bhotias, Mishmis, Mishis, Apatanis, Bodos, Nagas, Santals and the Chakmas. The international boundary between India and Myanmar does not serve as an ethnic divide. The different sections of the Naga population, Manipuri tribes, like the Hmars, Anals, Kukis, Thadous, Riangs, Vaiphei, Kabui, Kacha, Mao, and Tangkhul live on either side of Indo-Myanmar border. It would be difficult to superimpose an international boundary line on their traditional territories. They were not conscious of this boundary when Burma was one of the provinces of the British Empire. Several theories have been advanced to explain the concentration of the tribes in the hilly, forested and the dry regions of the subcontinent. The evidence, however, suggests that they were moved out of the river valleys by the peasants. The regions of tribal concentration are negative to the agriculture pursuits by and large. We have highlighted the possibility of demarcating the ethno-lingual regions based on the criteria of the race and dialect. As already noted the subcontinent has a baffling diversity in population composition, physical types, physical appearance, food and dressing preferences. The story is not complete yet. There is diversity in belief systems and the higher cultures. We know that hundreds of dialects are spoken. There are at least 23 languages, which have an established literary tradition. The *Janajatis* and the *Jatis* make up the bulk of population of the subcontinent. Dialect/language and culture act as a binding force because the people identify themselves with certain regions, natural habitats and ethnic features.

It is assumed that the political system has its roots in the socio-cultural and ethnic foundations. The religion of the people is basically Hinduism with underpinnings of Buddhist and Jaina tradition. The base is constituted by the Hindu polity imbibed by the extra-Indic religions, such as Judaism, Christianity and Islam. The Hindu view is that the *raja* or the *maharaja*, are incarnations of the divinity. This tradition is so constructed that the kings and their subjects have lived in mutual harmony without any basis for revolt or confrontation. At the same time, there is no scope for any revolutionary change. Even today the people of the subcontinent have faith in the scions of the princely houses all over the country. In the 55 years of the parliamentary democracy people's voting behaviour has always been tilted towards the candidate of the princely houses. This is true for the Rajputana states or states in Kathiawar, Bombay presidency, Orissa and other parts of the peninsular India.

The Road Ahead

I do not believe that the people of the subcontinent have an uncertain future. It is as uncertain as the future of the rest of the world community. We may say that the pot is boiling. New dispensations are being offered for the old ailments. New demands for the devolution of power to the people at the grassroots level are being orchestrated and entertained. In many cases the public demand for the new states (provinces) is being willy-nilly supported by the combination of political parties holding power at the centre. The centre here may mean New Delhi, Islamabad, Dhaka or Colombo. The people of Nepal are being exposed to a new political ethos in which the masses, not the monarchy of the common people, men and women, will rule over the country. Likewise, both the Sinhalese and the Tamils are working for a truce, which is evasive. New state structures are emerging out of the old and the ageing states. Chhattisgarh out of Madhya Pradesh, Jharkhand out of Bihar and Uttarakhand out of Uttar Pradesh are half-heartedly agreed solutions to the problem of emergent regionalism. Parliamentary elections have been contested and won on the issue of creating a new state in Andhra Pradesh, e.g. the Telengana state. If Jharkhand is there, Telengana and Vidarbha cannot be far behind. Indian federalism is on test. Unlike India, Pakistan, Bangladesh, Sri Lanka and Nepal have a fragile polity. In many situations military is holding power and they are doing so in the so-called integration of the state.

The new generation of citizens have their own perceptions about the general elections and working of the parliamentary democracy. They have their own ideas as how to manage these institutions in the light of the changing perspectives on the digitised knowledge oozing out of the computers. They enable them to browse through four hundred years of history in forty minutes. In what perspective do they draw inferences? The political development today will determine their future. Today is dependent on yesterday and tomorrow is still going to take shape. Their major problem is how to correlate the ethno-lingual regionalism with the on-going process of development as expressed in urbanisation trends, industrialisation and education. What are the implications of one omnibus model of development? How to ensure that the gains of development reach the masses and are not siphoned off appropriated by the more dynamic sections of population?

Time has come for the South Asian nations to organise themselves into a kind of 'soft' union, say on the pattern of European Union (EU). As a first step towards this unification, the existing political boundaries will be respected and retained as such, but significant changes will take place in the state-to-state relations as well as their collective relations with the rest of the world. Conditions should be created so that travellers from other parts of the world outside South Asia may travel on one visa. This would enable them to travel to any country or countries of their choice. The people of South Asia may be allowed to travel with a passport, or an identity card without a visa. The identity card should include information on their origin, profession/occupation and the purpose of their visit without a visa. They may apply for a Permanent Residence (PR) status in case they want to establish their business in the country of their choice. According to this plan, goods and services would move freely from one country to the other; tariff regulations will be common to all countries. This would mean that the South Asian nations would have a common tariff rates policy. These remarks may appear utopian. One may argue that the time has not come for such drastic changes. Eventually one single currency will be adopted like the Euro-Dollar. Exchange rates will be determined by the export-import regulations governing the bilateral/multilateral trade. A single organisation, such as the World Bank or the Asian Development Bank (ADB) would monitor the inward/outward flow of goods from one country to the other.

Political disputes between the countries will be sorted out on the basis of historical practices and contemporary realities. For example, vexed territorial problems, such as Jammu and Kashmir, will need fresh assessment. The new agenda will be to reorganise territory in order to eliminate the chances of military conflict, bilateral or multilateral. India and Pakistan will withdraw their forces from Kashmir. The two countries will govern the state as a joint partnership, with no ceasefires and no lines of armistice. The continuing conditions of governance will be retained as such. But situations, such as in Afghanistan will call for a new mindset. US hegemony can be encountered collectively. The other South Asian nations cannot be just spectators. The model of dictatorship and democracy as in Pakistan and Bangladesh should not be allowed to continue. But in case democracy is not a successful solution of the problems in Pakistan, a guided dictatorship like that of the Musharraf regime should be welcomed. India is a traditional democracy, even though national integration is a far cry. Alienating the religious and linguistic minorities cannot be the objective of the state. In a multi-ethnic state, minorities will have to be recognised and respected. The minorities will be required to come up to the expectations of the majority. The minority syndrome has to be overcome by committing to the larger cause of the national integration and multi-lateral development. Bangladesh will retain her political and sovereign status, but it will open the borders, and enter into an agreement with the Indian state of West Bengal to experiment with the idea of a Greater Bangladesh. Nepal was a functioning monarchy. But there was a rupture between the royalty and the popular movement aimed at overthrowing the monarchy and transfer of power to the people. Similarly, Sri Lanka is a case of democracy with an ideal two-party system. However, the Tamil sensibilities have to be respected within the democratic set-up of the island nation, avoiding the partition of the country.

In this agenda of reorganisation of the states it would be necessary to have a common defence force pooling together the existing armed forces and committing them to the defence of the entire subcontinent. For example, a representative body should be created on the lines of North Atlantic Treaty Organisation (NATO) which will take care of the region as a whole. An obvious implication would be to recognise the need for the collective security. Under the new political framework a collective responsibility would be necessary. An attack on Afghanistan or Pakistan will be treated as an attack on South Asia. It will be necessary to recognise the need for collective defence system including the defence resources and nuclear research findings will be shared for the common good of the people. The present expenditure on the assembly of nukes will be directed to human development sector and education. There will be no scope for the arms race, as the nations will work towards the higher goals of human development.

The present system of parliamentary democracy will be retained. Parties or individuals will make an effort to gear the democracies to fight successfully the tendency of dictatorship. In order to achieve these goals a new South Asian Consultative Committee will be created. The elected representatives will constitute it. This Committee will be the highest consultative body for monitoring economic development and for evolving a consensus on the basic issues of political identity and regional development. The Committee will meet four times in a year turn by turn at Kabul, New Delhi, Islamabad, Dhaka, Colombo and Kathmandu. A serious dialogue among the nations and between the nations will replace the SARC for good relations among the states. The armed forces will be reduced to one-third of the present strength. Expenditure on defence, such as armaments, ammunition and personnel will be reduced drastically. The combat planes, warships, choppers and the rocketry along with inter-continental ballistic missiles (ICBM) will be given to the custody of the South Asian Security Council. In this environment defence budget will shrink phenomenally. This will enable the sub-continental nations to divert the funds to sectors of human development, such as education, health, housing and employment.

All monuments from Taj Mahal, Harappa, Ajanta, Elora and the temples and mosques of historic importance will be given to the charge of a single department of archaeology to be designated as the Archaeological Survey of the subcontinent. If Europe can live with peace why can't we?

A Prognostication of the Future

The following time horizons for changing the present in order to allow the future to happen are being suggested below:

A. The Subcontinent by 2020

(i) Bangladesh will merge with the Indian state of West Bengal towards the formation of the Greater Bangladesh.
(ii) Kashmir is likely to be granted an autonomous status with the merger of the territory held by India and Pakistan. Both Laddakh and Jammu will be an integral part of the new sovereign state. India and Pakistan will collaborate in the development programmes of South Asia.
(iii) The present political construction of the Indian space will undergo a drastic change. Moving in and moving out of the Indian Union will be voluntary. There will be a set of guiding principles for the reorganisation of the states. The central government will be managed and run by a combination of regional parties. Thus the initiative will go to the hands of the regional parties. This will present a model of controlling the centre by the periphery.
(iv) Oriya-Bengali and Assamese language groups will be members of a dominion for the development of these languages and the cultural attributes. Parts of northern Bihar where the Maithli speakers are concentrated will go along with this new formation.
(v) The stage will be set for the eventual rise of Maharashtra as a quasi sovereign state. This will be completed in phases. Maharashtra may also try to have more autonomy. Initially Gujarat will go along with it because of the trade links between the two states. But the historical Maratha state will be constituted. It is likely that the Maharashtra state will incorporate part of Madhya Pradesh to become a multi-cellular state. It will be controlled from three centres of power located in Pune, Indore, and Gwalior. This would mean transfer of a part of Bundelkhand from Uttar Pradesh to the newly created Maratha state.

B. The Subcontinent by 2050

(i) The political restructuring as visualised by 2020 will inspire new combinations to emerge out of the old states. This process would be completed in several phases. The new state of Yadav land or Yadav Rajya will emerge out of parts of eastern Uttar Pradesh and Bihar, thus reorganising further the political space of Uttar Pradesh.
(ii) Central Uttar Pradesh including the districts lying east of the Ganga to

the west of Ghaghara will unite to constitute the new Rajya of Uttara Kaushal. It will be ruled from Ayodhya and Varanasi, which will be the second capital city. The pre-eminence of Lucknow will diminish finally.

(iii) The region presently known as Haryana and the western districts of Uttar Pradesh from Saharanpur to Mainpuri and Farrukhabad will unite to constitute the Jat Rajya.

(iv) The idea of greater Tamil Nad will be implemented in parts. It will subsume the Tamils of Sri Lanka and enter into a treaty with Kerala and Karnataka to share the river waters of Kaveri and Periyar.

(v) The newly created state of Chhattisgarh will merge with Jharkhand to constitute a tribal state in mid-India to become a part of the single tribal state. This would mean reorganisation of the three states of Gujarat, Rajasthan and Madhya Pradesh. The three new states will be created as follows: Bhilwara, Gondwana and the Santhali Pradesh. Initially the three tribal states will collectively evolve policies and programmes of regional development by sharing their natural resources. The basic unifying principle will be the geographical distribution and concentration of *janjatis*. Bhils will be the first ranking group in Bhilwara, Gonds in Gondwana and Santhals in the Santhali Pradesh.

(vi) The sub-Himalayan tribes from Laddakh to Arunachal Pradesh and Nagaland consisting of Laddakhis, Gujjar Bakarwals, Gaddis, Nepalese tribes, Bhutias, Bodos, Nishis and Apatanis to mention only a few will find it more rational to unite themselves in the sub-Himalayan tribal states at the highest order of regionalism. These groups will then organise themselves into several clusters within the state on the pattern of Gorkha Autonomous Council within West Bengal. Nagas from Nagaland and Manipur to Mishmis, Lushais, Mizos, Chakmas, Garos, Khasis and Jaintias will have new political formation on the southern and the eastern margins of the Assam valley. There will be a Development Council for each *janjati* with a distinctive ethnic character. The council will consider, approve and monitor the progress of development projects on a larger scale than the present situation permits. It will be possible to achieve this goal in the light of the principles of integrated regional and sectoral planning. The task will be assigned to a tribal development bank (TDB). The bank will manage the funds, and issue guidelines for the completion of development projects based on its funding. The tribal state will pool their resources available to them in the form of natural endowments, human skills and suitable technologies.

Appendix I

Nepalese Geographers' Perceptions

J.B. Burathoki (1924-74)

'Nepal is a laboratory of the altitude geography. Nepal is the only country, which presents a unique example in the world that within less than 90 miles from 300 feet or less above the Mean Sea Level the highest point on the earth, the Sagar Matha, rises to 29,028 feet. Because of the altitude factor within such a small country, except for the equatorial and the hot desert types, practically all the possible types of climates are found in Nepal. The rich wild life resources in response to diverse climatic conditions are found in a glaring example of bio-diversity.'

'From the point of view of human ecology most interesting study yet to be made of scores of different colourful communities, their social customs and dialects, etc., includes the Tharus, the Rajhabansis, the Kusundas, the Bhujels, the Sunwars, the Rais, the Limbus, the Sherpas, the Lepchas, the Gurungs, the Magars, the Tamangs, the Newaris, the Jumlis, the Humlis, the Manangis, the Mustangis and so on'.

'From the geographical and cultural point of view it can be said that Nepalese landuse pattern is diverse—the *terai* land is the agricultural backbone of the country. On the other hand, around 74 per cent of highland and mountainous area of the country has a variety of agricultural and pastoral type of land use, which has come under tremendous pressure in recent decades. The result is that many remote areas have a subsistence type of economy. The Khoriya, a form of shifting agriculture in rotation of 5 or 6 or 7 years in high mountains is very fast. Some of the problems may be highlighted—rapid deforestation, soil erosion and land slides, preservation of the forest ecosystem, unregulated aforestation in the *terai* as well as in the hills. If these steps are not taken seriously the fascinatingly picturesque Himalayan kingdom will become the Nepal Himalayan desert' (p. 31).

Sharma Hari Shrestha (1931-)

'While ecologists can contribute much...to achieve enhanced agricultural productivity that ultimately will lead to higher standard of living of the people and the general welfare of the nation as a whole. As geography subject has been made an optional paper at the secondary level of education in Nepal, the number of geography students is (has been) diminishing' (p. 34).

H.B. Gurung (1939-)

'The future development strategy—the basic approach is one of privatisation including foreign investment. I am unable to resolve the inherent contradiction between such an open economy and democratic socialism in a country of absolute poverty.'

'There never was a magic to development, least of all for an accelerated one.'

'The Hindukush Himalaya is a region of environmental complexity. The region is a geographical fact and an inherent reality. Complexity is not a problem to be solved but a phenomenon to be admired and enquired. But if you are referring to the environmental problem.... can be resolved only partly, pertaining to these due to human pressure by alleviating their poverty to counter pressure' (p. 40).

U.M. Malla (1932-)

'Land, climate, water, forest, soils—all are components of nature and naturally people tend to identify environmental deterioration as the consequence of natural factors... It may not be very correct in saying that social factors cause environmental problems... You are right if you say that social problems have definitely and hugely aggravated the environmental problems to such a distracting extent. Since we can do little to avoid natural occurrences, excessive rain, accumulation of water in the glacial lakes and incessant pull of gravity in the mountains, no doubt we have to solve the problems by minimising the harm caused by human beings through their number, behaviour, ignorance and poverty. Land degradation is beset by a fundamental theoretical confusion... Facts, ideologies and beliefs are yet far from being identified and relevance and accuracy of much of the data base still remain in doubt... Hence along with the introduction of various strategies of providing long perspectives a sustainable natural resource management to attain the basic conservation objectives in a way that reflects the social and cultural values and the economic needs of the people, there should be a concerted effort to bring about the attitudinal change among the people so the geographers should come forward in their expertise in multi-disciplinary approach in mitigating, if not eradicating, the environmental problems by the combined use of natural and social remedies.'

'Planning based on geography is more realistic and 'earth bound'; geographers can avoid building castles in the air and bring about a balanced development as they take into account the spatial variation in patterns of distribution of various physical, financial, cultural and human resources. A welfare society needs better allocation of commodities, better distribution of commodities and better allocation of means of production among the people and the places—all these are possible if geographers who deal with man-environment interaction and examine the spatial distribution of phenomena are actively involved in the process of planning and in the formulation of public policies at the local, regional, national and international levels' (p. 56).

C.B. Shrestha (1934-)

'Altitude geography is a part of geographical analysis. Such analysis would be welcomed as relevant in the context of mountainous countries like Nepal. However, the treatment of altitude geography as a specialised field is still a problem owing to inadequate theoretical base. All disciplines build upon their past, but they also change direction and emphasis. Geography will also develop beyond what previous geographers have done. However, the core of geographical enquiry will not change...The complexity of World Organisation in future will increase considerably as a result of reduction of economic locational constraints and tyranny of distance. This will create greater areal differentiation than has ever existed before due to the dominant role of human preference. This will call for more analytical approach. Therefore, geography in future needs to be more explanatory, predictive and prescriptive. It should be a more future-oriented science. Geography in future should change its focus from why and where things are to where things can and should be. In order to deal with complex situations, geographers will have to develop new creative skills and better theories and models' (p. 70).

Source: Panday, R.K. (1998).

Appendix II

The Regional Concept

Do the regions exist in the minds of geographers or are they real and can be recognized and delimited? Does it mean that the task of geographers is to identify regions and evolve criteria to define their personality? It is common knowledge that regions are identified on the basis of given characteristics of a part of the earth, or a country or a region within the same country. Such regions are called formal regions. For example, regions based on physiographic criteria are identified on the basis of their dominant relief features as expressed on the surface. The best example is the Ganga valley. In the recognition of the river basins it is absolutely necessary to carve out regions on a solid basis, taking the principle of the catchment area of the river as the basis. The valley of Ganga, on the basis of this criterion should include the valleys/basins of all the tributary rivers of the mainstream. The Ganga's alluvial plain is marked on the land but the concept of the catchment area will have to take into account small and large streams joining it on its right bank and originating in the Peninsular plateau. The principle of water-divide, therefore, emerges as a critical criterion in delimiting the regions. All the tributary streams, which are joining the Ganga, including the Yamuna and its tributaries, such as the Chambal and Banas belong to this region. The region also includes the basins of other rivers, such as Betwa, Ken, and Sone. Thus while the Ganga plain remains confined to the central segment of the North Indian Plain, the Ganga valley will have to include all the basins of the rivers joining the Ganga from the peninsular plateau. The source of the Chambal, for example, lies in the far-off region of the Vindhyan plateau near Mhou in Madhya Pradesh. If we are trying to draw a line on the basis of water-divide principle, the delimitation of the valley would incorporate vast chunks of the Peninsular plateau which slopes in the north-northeast direction. This also means that the Indo-Gangetic divide lies in the Aravalli Range and extends up to the Yamuna in Delhi and Haryana.

Similarly, formal regions can be recognized on the basis of political/ administrative structure. The problem gets further complicated when the concepts are refined and we try to delimit regions on the basis of climatic criteria. The climatic regions are delimited on the basis of meteorological parameters. Climatologists, such as Koeppen and Thornthwaite used multiple criteria in evolving a scheme of climatic regions. Koeppen considered temperature, amount and periodicity of rainfall/precipitation as the basis for climatic regionalisation. On the other hand, Thornthwaite considered temperature efficiency, rainfall effectiveness and the principle of water balance for delimiting climatic regions.

Appendix III

Racial Composition of Population: Comparative Ethnography

Author	*Racial Types*	*Region*	*Characteristics*
Denzil Ibbetson & J.C. Nesfield* Herbert Risley (1891)**	Turko-Iranian	Afghanistan, parts of Iran, Baluchistan, North Western Frontier Province (NWFP) and Kashmir.	
	Indo-Aryan$	Afghanistan, Kashmir, Punjab, Rajputana (now Rajasthan).	Dolicho-Cephalic, narrow-nosed.
	Mongoloid	Himalayas, Gilgit, Hunza, Hazara, Ladakh, Nepal, Assam, Burma (now Myanmar).	
	Dravidian@	Ceylon (now Sri Lanka), Peninsular Plateau to the Ganga Valley, Madras Presidency, Hyderabad, Central Provinces, Central Indian Princely States up to Chhattisgarh and Chotanagpur Plateau, parts of lower Bengal.	Dolicho-Cephalic with 'very broad nose.'
	Aryo-Dravidian	United Provinces of Agra and Oudh, Bihar, Parts of Rajputana (now Rajasthan), Ceylon.	Long head-form with a tendency to medium.
	Mongolo-Dravidian	Lower Bengal (including Bangladesh, West Bengal and Orissa).	Broad head (Brachy-Cephalic), Nose medium, with a tendency to broad.
	Scytho-Dravidian	Western India, Pakistan, parts of Baluchistan, North Western Frontier Province of Pakistan, undivided Punjab (including Haryana), Rajasthan.	Head is broad, nose moderately fine.
B.S. Guha (1931, 1944)	Negrito Type	Hills of southwestern India, Rajmahal Hills.	Perambiculum similar to Melanesian pigmies; Hair separated into locks, hair texture fine, downy or woolly nature, peppercorn hair, a variation of the Negroid stock, pigmy stature, small head, bulbous forehead, deeply pigmented, round, medium and long head, arms longer than legs.
	Proto-Australoid Type	Southern Peninsular Plateau from Godavari to Kanyakumari along the Eastern Ghats, Upper Narmada-Tapi River Basins and Chhattisgarh Plateau, Chotanagpur Plateau up to the Rajmahal Hills.	Representatives in aboriginal population, Peninsular India, similarity with Australian tribes, hence Proto-Australoids, representatives in Chenchus, Malayans, Kurumbas and Yeruva, south India, elements also found among Munda, Santals and Kol tribes.

(Contd.)

Author	Racial Types	Region	Characteristics
	Mongoloid Type	Himalayas from Gilgit, Hunza, Baltistan, Ladakh to Arunachal Pradesh including Nepal, Sikkim, Bhutan, represented among tribes in Assam, Indo-Burmese borderlands; another variety: short, broad headed, darker skin, rounder face represented, Burma, Chittagong hill tribes, Chakma and Magh; another variety: Tibeto-Mongoloid type, similar characteristics, tall stature, light skin colour, face long and flattish, massive head, Nepal, Sikkim and Bhutan infiltrated from Tibet.	Scanty growth of hair on the face and body, flatness of the facial parts, prominent cheek bones, peculiar shape of the eye, a loose fold of skin covering the inner angle, half open slit-like eyes; early branch of this type described as Palae-Mongoloids, medium stature, dark to light brown skin colour, short face, medium nose of low elevation, head varies from long to medium.
	Mediterranean Type	Gujarat and the lower basins of Narmada and Tapi, dominant type among the Dravidian speaking people of Peninsular India, Friharr von Eickstedt described as Melanid race; Guha does not agree with it; the mainstream of the Mediterranean race reported to be the builders of the Indus civilisation. But the Aryan speaking Vedic invaders pushed them into the Gangetic basin, some streams went beyond the Vindhyas, recognised as a dominant group in the population of North India, came from Asia minor and Arabian Peninsula, representatives found in Punjab where their presence strongest; also found in Sind, Rajputana and western U.P.	An early branch is generally referred to as Palaeo-Mediterranean: medium stature, dark skin, slight build, long and narrow head, small nose, fleshy and moderately broad, scanty development of hair. Tall stature, light skin colour, varying from dark to light olive brown, long head, arched-forehead, long face, developed chin, hair and eye colours dark brownish to black, large open eyes, people in Uttar Pradesh, Bombay, Bengal presidencies across the Malabar coast. A late stream of the Mediterraneans recognised as the Oriental Group. Very close to the main Mediterraneans; different from them; nose long, often convexed.
	Western Brachycephals	Lower Bengal along with Bangladesh, three main streams recognised: Alpines, Dinarics, and Armenoids (Dalmatia to Croatia), these types drifted along the western littoral region from south Baluchistan through Sind, Kathiawar, Gujarat, Maharashtra into Kannada, Tamil Nad, and then into Ceylon. An eastward movement drifted into the Gangetic delta, its trail in Central India, Eastern U.P. and Bihar. A subsidiary drift of the Dinaric race entered the Northwestern Himalayas and moved into the Western Himalayas and Nepal. The type present all along region from the Kho and the Burusho of Chitral and Gilgit; intermixing with the Oriental race. Common among the people of Bengal and Orissa; intermixing with the Mediterranean type; Kathiawar, Kannada, Tamil Nad, purest form in Coorg, broad heads are predominant in Gujarat, Gujaratis have a shorter face, indicates great admixture with the Alpenoid groups.	Alpines: medium stature, short, wide head, rounded occiput, round face, prominent nose, light complexion. Dinarics: taller stature, darker skin colour, receding forehead, high cranial vault, long nose, more often convexed. Armenoids: carry some of the main characteristics of the earlier two types.

Author	Racial Types	Region	Characteristics
	Nordic	Afghanistan, NWFP, undivided Punjab, parts of western U.P., parts of northern Rajasthan.	Powerfully built body, tall stature, long massive head, arched forehead, protruding occiput, long face, strong chin, light brown, white hazel, grey, green and blue eyes, narrow long nose, flesh coloured skin, gold, blond, blue hair and eyes; original racial traits were modified under tropical conditions of north-western India; multiple hereditary factors, genetic changes/mutations further modified these traits.
Giuffrida-Ruggeri	Negrito	No representatives so far discovered, comparisons with Bhils and Gonds, found in Central India from Gujarat, Madhya Pradesh, Rajasthan, Andhra Pradesh and Orissa.	
	Pre-Dravidian (Austroloid-Veddaic)	Central India from Gujarat to Orissa, Chhatisgarh, some tribes of Jharkhand and Rajmahal hills.	
	Dravidian	Resemblance with Ethiopians, excluding Somalis and Gallas, Mesati-Cephalic and the Brachy-Cephalic peoples of the Kanarese-Telugu regions, according to this view Dravidians came to India from Iran.	The Cephalic index among the inhabitants of the south having mixed with Brachy-Cephalic Negritos, tall Dolicho-Cephalic elements found among the Toda tribe.
	Dolicho-Cephalic Aryans	Include Kafirs, Dards, Rajputs, and the Kanets of Lahaul Himachal Pradesh.	Chanda's hypotheses of emigration of the Alpine folk; Ghurye finds it untenable.
	Brachy-Cephalic Lucoderms		
G.S. Ghurye	The Indo-Aryan Type	Confined to the undivided Punjab, Rajputana, and parts of the United Provinces; mixed with the Pre-Dravida and Munda types.	
	Pre-Dravida Type	Jungle folks of south India, a representative of this type.	
	Dravida Type	Mixed with the population of the Tamil and Malyalam region.	
	The Western Type	Western coast upto the northern limits of malabar, Mysore, and the Telegue country and the whole of Maharashtra.	
	Munda Type	Around Chotanagpur Plateau (Chhattisgarh, Jharkhand and Highland Orissa), Bihar (formed by the mixture of Indo-Aryan, Pre-Dravida and the Munda); Bengal and Orissa, a combination of western, Munda and the Mongoloid types.	
	Mongoloid Type	The Mongoloid type confined to the Himalayas, Nepal, Sikkim, Bhutan and Assam.	

Notes:

(i) * J.C. Nesfield (1855): *Brief View of the Caste System of the North Western Provinces and Oudh.*

(ii) ** Sir H.H. Risley (1891): *Tribes and Castes of Bengal, Ethnographic Glossary* (in two volumes). Also, Anthropometric data for Bengal. ibid. (1915): *The People of India* (Second Edition).

(iii) @ The term Dravidian implies language, not the people. But in early writings on ethnography the term has been extensively used to imply people (read Dravidian speaking people); 'possibly the original people of India' (Ghurye: *Caste and Race in India,* fifth edition, 1969, p. 126). The type is supposed to have spread over a vast region from Ceylon to the Valley of Ganges, best representatives being Nayars, Parayans, Santals and others. 'I have, therefore, to connect the Santal with the Jungle tribe, and separate the Nayar and others like him from them' (Ghurye: Op. cit. pp. 128-29).

(iv) $ The term used to denote Indo-Aryan proposed by Risley is 'unsatisfactory' (Ghurye: Op. cit. p. 126). The mixutre of Indians and the Aryans is not acceptable because this would mean mixture of Indians with Aryans or Iranians; 'I shall retain the term Indo-Aryan to denote the Aryan type of India' (ibid. p. 126). 'By Indo-Aryan... I mean the physical type to which the Vedic Indians largely contributed' (ibid.). 'The physical type of the Punjab is so uniform as to preclude any possibility of large mixture' (ibid.)

Appendix IV

Cultural Succession in the Subcontinent, From Neolithic to Chalcolithic

Period/Region	*Sites*	*Time*	*Evidence*
1.0 Pre-Harappan 1.1 North-West 1.1.1 Baluchistan: (Pakistan-Iranian Borderlands)		Circa 4500-1200 B.C.	Food Producing Economy
	West of Indus: River Valleys of Zhob, Beji, Pishin, Lora, Dasht, Kel, Lora lai, Kili Gul Mohammad, Near Quetta Periano Ghundai, Rana Ghundai		Mother Goddesses, Zhob Goddesses, Peepal leaf motifs on pottery
	East of Indus: Damb Saadat, Togau, Siah Damb, Anjira, Nal, Amri		Nal and Amri wares, Cotta wares, Built-up houses
1.1.2 South-Eastern Afghanistan	Kot Diji	Circa 2500 B.C.	
	Mondigek		Grey and black burnished pottery, Mother Goddesses, hearth ashes beds
1.1.3 Kashmir		3000–2000 B.C.	
	Burzahom (located at Karewa Beds, near Pahalgam)	Second half of the second millennium B.C.	Grey and black burnished pottery remains, Stone axes, Stone and bone composite tools, Skeletal remains of dogs buried with their masters
1.2 West 1.2.1 Rajasthan		Circa 2500 B.C.–1800 B.C.	
	Kalibangan (located on the banks of the Saraswati river—now Ghaggar river)	1800 B.C. –1500 B.C.	Potsherds of northern polished wares with stone axes
1.3 Southern Deccan	Tapi Valley, Prakash, Bahal, Chandeli, Sonegaon, Burhanapur	Circa 2500–1500 B.C.	
	Krishna and Godavari River Valleys: Brahmagiri, Sangana Kallur, Piklihal, Maski, Tekhela Kote, Hallur, Utnur, Kupgal	First Phase: Circa 2500–1800 B.C.	Ground stone, Axes, Hand-made pottery, Grey ware pottery, Domesticated animals such as goats, sheep, cattle
		Second Phase: Circa 1800–1500 B.C.	Ground stone, Axes, Hand-made pottery, Grey ware pottery, Copper and bronze tools; Domesticated animals such as goats, sheep, cattle

Source: Table compiled by the author with the help of standard works on Pre-historic India.

Appendix V

Development Indicators

Sl. No.	Indicators	Year	Afghanistan	Bangladesh	Bhutan	India	Nepal	Pakistan	Sri Lanka	Maldives
			Countries							
A. Demographic										
1.	Total population (000)	2000	22,720	129,155	2,124	1,013,662	23,930	156,483	18,827	286
2.	Annual growth rate (%)	1985-2000	3.0	1.7	2.4	1.9	2.5	2.9	1.1	2.9
3.	Life expectancy at birth (years)	1995-2000								
	a. Male		45.0	58.1	59.5	62.3	57.6	62.9	70.9	65.7
	b. Female		46.0	58.2	62.0	62.9	57.1	65.1	75.4	63.3
4.	Level of urbanisation (%)	2000	21.9	24.5	7.1	28.4	11.9	37.0	23.6	26.1
5.	Urban population (000)	2000	4,971	31,665	152	288,283	2,884	57,968	4,435	75
6.	Annual growth rate of urban population (%)	2000-2015	5.3	3.7	5.8	2.7	4.9	3.9	3.1	3.8
7.	Rural population (000)	2000	17,749	97,490	1,972	725,379	21,087	98,515	14,392	211
8.	Annual growth rate of rural population (%)	2000-2015	2.5	0.6	2.2	0.5	1.6	1.2	0.2	2.1
B. Infrastructure										
9.	Water supply coverage (%)	2000								
	a. Total		13	97	62	88	81	88	83	100
	b. Rural		11	97	60	86	80	84	80	100
	c. Urban		19	99	86	92	85	96	91	100
10.	Sanitation coverage (%)	2000								
	a. Total		12	53	69	31	27	61	83	56
	b. Rural		8	44	70	14	20	42	80	41
	c. Urban		25	82	65	73	75	94	91	100
C. Energy										
11.	Commercial energy production (kg. of oil equivalent)	1998	—	21,495	—	386,180	6,355	40,128	4,022	—
12.	Commercial energy use (kg. of oil equivalent /capita)	1998	—	198.8	—	469.9	327.4	432.8	337.2	—

(Contd.)

Sl. No.	Indicators	Year	Afghanistan	Bangladesh	Bhutan	India	Nepal	Pakistan	Sri Lanka	Maldives
			Countries							
D. Land/Agriculture										
13.	Population density (persons per 000 hectare)	2000	348	9,922	452	3,409	1,673	2,030	2,913	9,533
14.	Land area (000 hectare)	1997	65,209	13,017	4,700	297,319	14,300	77,088	6,463	30
15.	Arable land (hectare per person)	1997	0.33	0.06	0.19	0.17	20.27	27.29	13.43	3.33
16.	Land use by type, irrigated land (% of crop land)	1997	34.77	44.81	25.00	33.56	38.24	81.39	31.78	—
17.	Permanent crop land (% of land area)	1997	0.22	2.50	0.43	2.66	0.49	0.73	15.78	6.67
E. Employment										
18.	Labour force	1998								
	a. Total (000)		10,521	64,071	364	431,056	10,512	48,685	8,075	108
	b. Women (%)		35.2	42.3	39.8	32.1	40.5	27.7	36.2	43.1
19.	GDP (per capita) (PPP US $)	1998	—	1,361	1,536	2,077	1,157	1,715	2,979	4,083
20.	Unemployment (% of labour force): Total	1995	—	—	—	—	—	5.4	12.5	—
F. Assets										
21.	Personal computers (per 000 people)	1998	—	—	4	3	—	4	—	—
22.	Telephone mainlines (per 000 people)	1998	1	3	16	22	8	19	28	71
23.	Television sets (per 000 people)	1995	10	7	17	61	3	88	78	40
G. Health, Education and Poverty										
24.	Hospital beds (per 000 People)	1995	—	—	—	—	0.2	—	—	2.0
25.	Health expenditure (% of GDP)	1998	—	3.5	6.9	—	5.5	3.9	2.57	10.7
26.	Illiteracy rate-population aged 15 years and over (%)	2000								
	a. Total		63.7	59.2	52.7	44.2	58.6	56.7	8.4	3.7
	b. Male		49.0	48.3	38.9	31.4	40.9	42.4	5.5	3.7
	c. Female		79.2	70.5	66.4	57.9	76.2	72.2	11.1	3.6
27.	Population below national poverty line (%)	1987-98	—	35.6	—	35.0	42.0	34.0	35.3	—

Source:

1. United Nations (1999): World Urbanisation Prospects– The 1999 Revision, United Nations, Department of Economic and Social Affairs, New York.
2. World Health Organisation (2000): Global Water Supply and Sanitation Assessment 2000 Report, World Health Organisation and United Nations Children's Fund, Geneva.
3. International Road Federation, World Road Statistics, World Bank, World Development Indicators, 2000.

4. United Nations World Urbanisation Prospects– The 1999 Revision, Food and Agriculture Organisation, Production Year Book, World Bank, World Development Indicators, 2000.
5. International Labour Organisation Data base, Estimates and Projections of the Economically Active Population, 1995-2010, International Labour Organisation Data base, Key Indicators of the Labour Market, 1999 (issue).
6. International Telecommunications United World Telecommunication Development Report 1999; World Health Organisation, World Health Report 2000; UNESCO Statistical Year Book, 1999; World Bank, World Development Indicators 2000.

Appendix VI

List of Scheduled Castes in Selected States of the Indian Union

Gujarat

Ager, Bakad or Bant, Bawa (Dhedh) or Dedh-Sadhu, Bhambi, Bhambhi, Asadru, Asodi (16 more castes), Bhangi or Rukhi, Mehtar, Olgana, Rukhi, Malkana (4 more castes), Chalvadi or Channyya, Chamadia, Chamar, Nalia, or Rohit, Chenna Dasar or Holaya Dasar, Dangashia, Dhor, Kakkayya or Kankayya, Garmatang, Garoda, Garoda or Garo, Hadi, Halleer, Halsar, Hulasvar, Holar or Valhar, Holaya or Holer, Lingader, Mahar, Taral or Dhegu Megu, Mahayavanshi, Dhed, Vankar or Maru Vankar, Mang, Matang or Minimadig, Mang-Garudi, Megwal or Menghvar, Meghwal, Mochi, Mukti, Nadia or Hadi, Pasi, Senva, Shemalia, Shenva, Chenva, Sedma or Ravat, Thori, Tirgar or Tirbanda, Turi, Turi-Barot, Turi-Barot or Dhedh or Antyaj.

Madhya Pradesh

Audhelia, Badri or Bagdi, Bahna or Bahana, Balahi, Balahi or Balai, Balai, Banchada, Bansphor or Basor, Barahar or Basod, Bargunda, Basor or Bansphor, Basor, Barud, Bansor or Bansodi, Bedia, Beldar, Beldar or Sunkar, Bhangi or Mehtar, Bhanumati, Chadar, Chamar, Ahirwar, Chamar Mangan, Mochi or Raidas, Bairwa, Bhambi, Jatav, Mochi or Regar, Chamari, Mochi-Nona, Rohidas, Ramnami, Satnami, Suriyabanshi, Jatav or Mochi, Chidar, Chitar, Dahati, Dahayat or Dahat, Dewar, Dhanuk, Dharkar, Balmik or Lalbegi, Dhed, Dhobi, Dohor, Dom, Dom or Dumar, Domar or Doris, Dome, Ganda or Gandi, Ghasi or Ghasia, Ghasia, Holiya, Kanjar, Katia or Patharia, Khangar, Kanera or Mirdha, Khatik, Chikwa or Chikvi, Koli or Katia, Koli or Kori, Kori, Kotwal, Kuchbandhia, Kumhar, Mahar, Mahar or Mehra, Mang, Dankhni-Mang, Mang-Mahashi, Mang Garudi, Madari, Garudi or Radhe-Mang, Mang or Mang Garodi, Meghwal, Mehtar or Bhangi, Mehtar, Bhangi or Dhanuk, Moghia, Muskhan, Nat, Kalbelia or Sapera, Pardhi, Pasi, Rujjhar, Sansi, Sansia or Bedia, Silawat, Zamral.

Maharashtra

Ager, Aray (Mala), Bakad or Bant, Balahi or Balai, Basor, Burud, Bansor or Bansoli, Beda (Budga), Jangam, Bedar, Bhambi, Bhambhi, Asadaru, Asodi (16 more castes), Bhangi, Mehtar, Olgana, Rukhi (5 more castes), Bindla, Byagara, Chalvadi or Channyya, Chamar, Chamari, Mochi, Nona (4 more castes), Chambhar, Chenna Dasar or Holaya Dasar, Dakkal (Dokkalwar), Dhor, Kakkayya or Kankayya, Dohor, Dom or Dumar, Ellamalwar (Yellammalawandlu), Ganda or Gandi, Garoda or Garo, Ghasi or Ghasia, Halsar, Haslar, Hulasvar or Halasvar, Holar or Holer, Holeya, Holaya or Holer, Holeya Dasari, Holia, Kaikadi, Katia or Patharia, Khangar, Kanera or Mirdha, Khatik, Chikwa or Chikvi, Kolupulvandlu, Kori, Lingader, Madgi, Madiga, Mahar or Mehra, Taral or Dhegu Megu, Mahyavanshi, Dhed, Vankar or Maru Vankar, Mala, Mala Dasari, Malajangam, Mala Masti, Mala Sale (Netkani), Mala Sanyasi, Mang, Dankhni Mang, Mang Mahashi, Madari, Garudi or Radhe-Mang, Mang Garodi, Matang or Minimadig, Manne, Mashti, Meghwal or Menghvar, Mehtar or Bhangi, Mitha Ayyalvar, Mochi, Mukri, Nadia or Hadi, Pasi, Sansi, Shenva, Chenva, Sedma or Ravat, Sindhollu (Chindollu), Trigar or Tribanda, Turi.

West Bengal

Bagdi or Duley, Bahelia, Baiti, Bantar, Bauri, Bediya, Beldar, Bhogta, Bhuimali, Bhuiya, Bhumij, Bind, Chamar, Charmakar, Mochi, Muchi, Rabidas, Ruidas or Rishi, Chaupal, Dabgar, Damai (Nepali), Dhoba or Dhobi, Doai, Dom or Dhangad, Dosadh or Dusadh, including Dhari or Dharhi, Ghasi, Gonrhi, Halalkhor, Hari, Mehtar or Bhangi, Jalia Kaibarrta, Jhalo Malo or Malo, Kadar, Kami (Nepali), Kandra, Kanjar, Kaora, Karenga or Koranga, Kaur, Keot or Keyot, Khaira, Khatik, Koch, Konai, Konwar, Kotal, Kuraiar, Lalbegi, Lohar, Mahar, Mal, Mallah, Mehtor, Musahar, Namsudra, Nat, Nuniya, Paliya, Pan or Sawasi, Pasi, Patni, Pod or Poundra, Rajbanshi, Rajwar, Sarki (Nepali), Sunri excluding Saha, Tiyar, Turi.

Orissa

Adi Andhra, Amant or Amat, Audhelia, Badaik, Bagheti or Baghuti, Bajikar, Bari, Bariki, Basor or Burud, Bauri, Bauti, Bavuri, Bedia or Bejia, Beldar, Bhata, Bhoi, Chachati, Chakali, Chamar, Mochi, Muchi or Satnami, Chandala, Cherua or Chhelia, Dandasi, Dewar, Dhanwar, Dhoba or Dhobi, Dom, Dombo or Duria Dom, Dosadha, Ganda, Ghantarghada or Ghantra, Ghasi or Ghasia, Ghogia, Ghusuria, Godagali, Godari, Godra, Gokha, Gorait or Korait, Haddi, Hadi or Hari, Irika, Jaggali, Kandra or Kandara, Karua, Katia, Kela, Khadala, Kodalo or Khodalo, Kori, Kuli, Kummari, Kurunga, Laban, Laheri, Madari, Madiga, Mahuria, Mala, Jhala, Malo or Zala, Mang, Mangan, Mehra or Mahar, Mehtar or Bhangi, Mewar, Mundapotta, Musahar, Nagarchi, Namasudra, Painda, Pamidi, Pan or Pano, Panchama, Panika, Pantani, Pap, Pasi, Patial, Patikar Patratanti or Patua, Rajna, Relli, Sabakhia, Samasi, Sanei, Sapari, Sauntia (Santia), Sidhria, Sinduria, Siyal, Tamadia, Tamudia, Tanla, Tiar or Tior, Turi, Ujia, Valamiki or Valmiki.

Andhra Pradesh

Adi Andhra, Adi Dravida, Anamuk, Aray (Mala), Arundhatiya, Arwa Mala, Bariki, Bavuri, Beda (Budga), Jangam, Bindla, Byagara, Chachati, Chalavadi, Chamar, Mochi or Muchi, Chambhar, Chandala, Dakkal (Dokkalwar), Dandasi, Dhor, Dom, Dombara, Paidi or Pano, Ellamalawar (Yellammalawandlu), Ghasi, Haddi or Relli Chachandi, Godagali, Godari, Gosangi, Holeya, Jaggali, Jambuvulu, Kolupulvandlu, Madasi Kuruva or Madari Kuruva, Madiga, Madiga Dau and Mashteen, Mahar, Mala, Mala Dasari, Mala Dasu, Mala Hannai, Mala Jangam, Mala Masti, Mala Sale (Metkani), Mala Sanyasi, Mang, Mang Garodi, Manne, Mashti, Matangi, Mehtar, Mitha Ayyalvar, Mundala, Paky or Moti, Pambada or Pambanda, Pamidi, Panchama or Pariah, Relli, Samagara, Samban, Sapru, Sindhollu (Chindollu), Thoti.

Assam

Bansphor, Bhuinmali or Mali, Brittial Bania or Bania, Dhupi or Dhobi, Dugla or Dholi, Hira, Jalkeot, Jhalo, Malo or Jhalo-Malo, Kaibarthaj Jalia, Lalbegi, Mahara, Mehtar or Bhangi, Muchi or Rishi, Namsudra, Patni, Sutradhar.

Arunachal Pradesh

Bania, Dholi, Hira, Kaibarta or Jaliya, Muchi, Namasudra.

Himachal Pradesh

Ad-Dharmi, Badhi or Nagalu, Balmiki, Chura or Bhangi, Bandhela, Bangali, Banjara, Bansi, Barad, Barar, Burar or Bawaria, Bazigar, Bhanjra, Chamar, Jatia Chamar, Mochi, Ramdasi, Ravidasi or Ramdasia, Raigar, Chanal, Chhimbe (Dhobi), Chuhre, Dagi, Daole, Darai or Daryai, Darain, Daule, Dhaki or Toori, Dhanak, Dhaogri or Dhuai, Dhogri, Dhangri or Siggi, Doom or Doomna, Dumna, Mahasha, Dumne (Bhanjre), Hali, Hesi, Jogi, Julahe, Kabirpanthi, Julaha or Keer, Kamoh or Dagoli, Karoack, Khatik, Koli or Kori, Lohar, Mazhabi, Megh, Nat, Od, Pasi, Pherera, Rehar, Rehara, Sanhai, Sanhal, Sansi, Bhedkut or Manesh, Sansoi, Sapela, Sarde, Sarare or Siryare, Sarehde, Sarera, Sikligar, Sipi, Sirkiband, Teli, Thathiar or Thathera.

(Note: Annendix VI illustrates the fact that the population is highly diversified. This diversity is due to the heterogeneous sections of population concentrated in enclaves in different parts of the subcontinent, particularly in the Indian union. While these different groups are distributed in mutually exclusive enclaves, in certain cases they are overlapping too. In urban and semi-urban parts of the subcontinent, tribes are generally missing. If one selects the district as a self-contained unit of space, the number of ethno-lingual groups is formidable. This means that they display a high degree of social diversity).

Select Bibliography

Abayasekara, A.W.A.D.G. (1979): *Population of Sri Lanka—A Reader*. Colombo: University of Colombo, Demographic Training and Research Unit.

Abbott, J.E. (1899): 'The Topographical List of Bhagvat Purana', *Indian Antiquary*.

Abhayaratne, O.E.R.and C.H.S. Jayawardene (1965): 'Internal Migration in Ceylon', *The Ceylon Journal of Historical and Social Studies*, 8(4): 69-90.

Afzal, M. (1981): 'Urbanisation and Internal Migration in South Asia - Some Facts and Data Sources', in M.I. Siddiqi (ed.), *Population Redistribution in Pakistan*, pp. 31-50, Karachi: University of Karachi.

Aghai, M.A. (1980): 'Population Policy and Programme', in M.I. Siddiqi (ed.), *Population Redistribution in Pakistan*, pp. 56-67, Karachi: University of Karachi.

Ahmad, Aijazuddin (1984): 'Social Geography', in Manzoor Alam (ed.), *A Survey of Research in Geography 1972-75*, pp. 67-84, New Delhi: Indian Council of Social Science Research.

Ahmad, Aijazuddin (1980): 'Integrated Rural Development in India—Social Implications of Planning since Independence', proceedings of the *Second International Symposium of Asian Studies*, Asian Research Service, Vol. II, Hong Kong. pp. 905-922.

_______________(1985): 'Regional Development Process and Redistribution of Tribal Population in Mid-India', in Leszek A. Kosinski and K. Maudood Elahi (eds.), *Population Redistribution and Development in South Asia*, pp. 65-78, Dordrecht: D. Reidel Publishing Company.

_______________(1995): 'Tibetan Immigrants in India', *Population Geography*, 17 (1-2): 7-14.

_______________(1993): 'Social Structure and Regional Development: An Introduction', in Aijazuddin Ahmad (ed.), *Social Structure and Regional Development: A Social Geography Perspective*, pp. 1-16, Jaipur: Rawat Publications.

_______________(1997): 'Ecology, Regional Differentiation and Cultural Synthesis: Rethinking India's Cultural Pluralism', in Rasheeduddin Khan (ed.), *Rethinking Indian Federalism*, pp. 56-72, Delhi: Raplica Press Pvt. Ltd.

_______________(1999): *Social Geography*, Jaipur: Rawat Publications.

_______________(2003): 'Demographic Fears are Baseless', *Indian Currents*, XV (27): 20-21.

Ahmad, Kazi S. (1955): 'A Geographical Study of the Refugee Population and Some of its Problems', *Pakistan Geographical Review*, 10 (2): 1-18.

Ahmad and Lees (ed; 1864-69): *Badauni's Muntakhbat-ut-Tawarikh*, Calcutta: Royal Asiatic Society of Bengal.

Ahmad, N. (1939): 'Al-Biruni's Geography Of India', *Calcutta Geographical Review*, 5 (1).

_________(1947): *Muslim Contribution to Geography*, Lahore.

Ahmad, Qeyamuddin (ed.; 1992): *Al-Biruni's India*, Abridged Edition of Dr. Edward C. Sachau's English Translation. New Delhi: National Book Trust, India.

Ahmad, S.M. (1953): 'Al-Masudi's Contribution to Medieval Arab Geography', *Islamic Culture*.

__________(1956): 'Arab Geography', *The Geographer*, Aligarh.

Alam, J. (1979): 'Mujhe Wapas Lo—Pakistanis in Bangladesh', *Bichitra*, 7 (46), Dacca, April 20 (Newspaper).

Alavi, S.M.Z. (1940): 'Arab Explorations in the 9th and 10th Centuries', *Islamic Culture*.

__________(1950): 'Survey of Arab Geographical Literature in the 9th and 10th Centuries A.D., *Islamic Literature*.

__________(1965): *Arab Geography in the 9th and 10th Centuries*, Aligarh Muslim University Publication.

__________(1966): *Geography in the Middle Ages*, Delhi.

Alchin, F.R. (1957): 'The Neolithic Stone Industry of North Karnataka Region', *Bulletin of the School of Oriental and African Studies*, 19 (2).

Ali, S.M. (1950): 'Some Geographical Ideas of Abu Rehan Al-Beruni', *The Geographer*, 2 (2).

________(1960): *Arab Geography*, Aligarh Muslim University Publication.

________(1966): *The Geography of the Puranas*, Delhi: New Age Publishing House.

Allchin, Bridget and Raymond (1968): *The Birth of Indian Civilization*. Harmondsworth: Penguin Books.

Ayoob, M. and L. Subrahmanyam (1972): *The Liberation War*, New Delhi: S. Chand Publishers.

Bardhan, A.B. (1976): *Tribal Problems in India*, New Delhi: Communist Party of India.

Basham, A.L. (1954): *The Wonder That Was India*, London: Oxford University Press.

__________(1964): *Studies in Indian History and Culture*, London: Oxford University Press.

Bennet, Charles Jr. (1980): 'The Morphology of Language Boundary: Indo-Aryan and Dravidian in Peninsular India', in David E. Sopher (ed.), *An Exploration of India*, pp. 234-251, London: Longman.

Berreman, Gerald (1960): 'Caste in India and the United States', *American Journal of Sociology*, 66: 120-27.

______________(1965): 'The Study of Caste Ranking in India', *Southwestern Journal of Anthropology*, 21: 115-29.

Beteille, Andre (1969): *Caste, Class and Power – Changing Patterns of Stratification in a Tanjore Village*, Bombay: Oxford University Press.

___________(1977): 'The Definition of Tribe', in Romesh Thapar (ed.), *Tribe, Caste and Religion*, Delhi: Macmillan Company of India Ltd.

___________(1977): 'Caste and Social Status', in Romesh Thapar (ed.), *Tribe, Caste and Religion in India*, pp. 59-65, Delhi: Macmillan Company of India Ltd.

___________(1992): *Society and Politics in India–Essays in a Comparative Perspective*. Delhi: Oxford University Press.

Beveridge, H. (1873): *Akbarnama*, Bibliotheca Indica, Calcutta.

Bharadwaj, S.M. (1973): *Hindu Places of Pilgrimage in India*, Berkeley: University of California Press.

Bhargava, P.L. (1956): *India in the Vedic Age*, Lucknow.

___________(1964): *The Geography of Rigvedic India*, Lucknow.

Bhatt, B.L. (1977): 'The Religious Geography of South Asia: Some Reflections', *The National Geographical Journal of India*, 23 (1-2): 26-39.

Bhattacharaya, N.N. (1991): 'The Contemporary Geo-Political Problems of North-East India', *North-Eastern Geographer*, 23 (1-2): 1-5.

Bird, J. (1822): 'On the Historical Geography of Hindustan and the Origin of the Social State among the Hindus', *Journal of the Asiatic Society of Bengal*, 9.

Bishop, R.N.W. (1952): *Unknown Nepal*, London: Luzac.

Bose, A.B. (1970): 'Problems of Educational Development of Scheduled Tribes', *Man in India*, 27 (1).

Bose, Ashish. (ed.; 1969): 'Some Aspects of Linguistic Demography of India', in *Language and Society of India*, pp. 37-51, Shimla: Indian Institute of Advanced Study.

___________, U.P. Sinha and R.P. Tyagi (ed.; 1990): *Demography of Tribal Development*, Delhi: B.R. Publishing Corporation.

Bose, N.K. (1971): 'Some Aspects of Indian Civilisation', *Man in India*, 51 (1): 1-15.

Bruce, C.G. (1934): *Himalayan Wanderer*, London: Maclehose.

Buttimer, A. (1969): 'Social Space in Interdisciplinary Perspective', *Geographical Review*, 54: 417-26.

Census of India (1901): Punjab and the North West Frontier Province, Report and Tables, 17a (II), Lahore: Civil and Military Gazette Press, 1902.

Census of India (1911): North West Frontier Province, Report and Tables, 13 (I-II), Peshawar: Anand and Sons, 1912.

Census of India (1921): North West Frontier Province, Report and Tables, 14 (I-II), Peshawar: Government Press, 1922.

Census of India (1931): North West Frontier Province, Report and Tables, 15 (I-II), Peshawar: Government Stationary and Printing, 1933.

Census of Pakistan (1951): North West Frontier Province and Frontier Regions, Report and Tables, 4, Karachi: Manager of Publications.

Census of Pakistan (1951): East Bengal, Karachi: Ministry of Home.

Census of Pakistan (1961): West Pakistan, 3, Karachi: Office of the Census Commissioner.

Census of Pakistan (1972): District Reports for NWFP Districts, 1 and 2, Islamabad and Karachi: Census Organisation, 1975-77.

Central Bank of Ceylon (1981): *Review of the Economy—1980*, Colombo.

Ceylon Department of National Planning (1962): *The Short Term Implementation Programme*, Colombo.

Ceylon Land Commission (1928): *Third Interim Report*, Sessional Paper II, Colombo: Government Printer.

Chakravarti, M. (1916): 'Notes on the Geography of Orissa in the Sixteenth Century', *Journal of the Asiatic Society of Bengal*, New Series 2.

Chakravarti, S.N. (1941): 'The Prehistoric Periods in India', *Journal of the University of Bombay*, 10 (1).

Chen, Lincoln C. (ed.; 1973): *Disaster in Bangladesh*, New York: Oxford University Press.

Christopher Von Fuerer-Haimendorf (1954): 'When, How and From Where Did the Dravidians Come to India?' *The Indo-Asian Culture*, 2 (3): 238-47.

Clark, W.E. (1919): 'Sakadwipa and Sveladwipa', *Journal of the American Oriental Society of Bengal Year Book*.

Cohn, B.C. and Milton Singer. (ed.; 1968): *Structure and Change in Indian Society*, Chicago: Aldine Publishing Company.

Cumberland, John H. (1971): *Regional Development Experiences and Prospects in the United States of America*, The Hague: Mouton.

Cunningham, Alexander (1975): *Ancient Geography of India*, Varanasi: Bhartiya Publishing House.

Cunningham, Alexander (1971): The Ancient Geography of India, *1*, *The Buddhist Period*. London: Roy.
Das, A.C. (1921): *Rigvedic India*, Calcutta.
Das, N. (1894): '*Geography of Asia Based on the Ramayana*', Journal of the Royal Asiatic Society.
Davis, K. (1951): *The Population of India and Pakistan*, Princeton: Princeton University Press.
De Silva, K.M. (1977): *Sri Lanka—A Survey*, London: Hurst Co.
De Tocqueville, Alexis (1959): *Journey to America*. London: Faber and Faber.
Demographic Training and Research Unit (1976): *Population Problems of Sri Lanka*, Colombo.
Desai, A.R. (1969): *Rural Sociology in India*, Bombay: Popular Prakashan.
Dev, B.J. and D.K. Lahiri (1981): 'Assam in the Days of Bhashani and League Politics', *Journal of the Asiatic Society of Bangladesh*, 25: 189-235.
Dey, N.L. (1927): *Geographical Dictionary of Ancient and Medieval India*, London.
Dhital, B.P. (1974): 'Population Growth and Agriculture', in D.C.J. Upadhyaya and J.V. Abueka (eds.), *Population and Development in Nepal*, Kathmandu: Centre for Economic Development and Administration.
Dias, H.D. (1978): 'Development Policy and Migration in Sri Lanka', *Sri Lanka Journal of Social Sciences*, 1 (1): 37-60.
Dikshitar, V.R.R. (1935): 'Geographical Data of the Deccan and South India as Gathered from the Ramayana', *Indian Culture*.
Dubey, B. (1967): *Geographical Concepts in Ancient India*, Varanasi.
Dumont, Louis. (1988): *Homo Hierarchicus—The Caste System and its Implications*, New Delhi: Oxford University Press.
Dutt, N.K. (1986): *Origin and Growth of Caste in India*, Calcutta: Firma Kim Pvt. Ltd.
ESCAP (1975): 'Comparative Study of Population Growth and Agricultural Change in Sri Lanka', Bangkok: Economic and Social Commission for Asia and the Pacific.
ESCAP (1976): 'Population of Sri Lanka', Country Monograph Series 4, Bangkok: Economic and Social Commission for Asia and the Pacific.
ESCAP (1980): 'Migration, Urbanisation and Development in Sri Lanka', Country Report 2, Bangkok: Economic and Social Commission for Asia and the Pacific.
Elahi, K. Maudood and Sabiha Sultana (1985): 'Population Redistribution and Settlement Change in South Asia', in Leszek A. Kosinski and K. Maudood Elahi (eds.), *Population Redistribution and Development in South Asia*, pp. 15-36, Dordrecht: D. Reidel Publishing Company.
Ellman, A.O. et al. (1976): *Land Settlement in Sri Lanka 1840-1975*. Colombo: Agrarian Research and Training Institute, Research Study Series 16.
Farmer, B.H. (1957): *Pioneer Peasant Colonisation in Ceylon*. London: Oxford University Press.
Farmer, B.H. (1974): *Agricultural Colonisation in India Since Independence*. London: Oxford University Press.
Fernando, D.F.S. (1980): 'The Continuing Fertility Decline in Sri Lanka', *Journal of Biosocial Science*, 12 (1): 51-60.
Fernando, T. (ed.; 1978): *Modern Sri Lanka: A Society in Transition*, Syracuse: Syracuse University.
Fleat, J.F. (1899): 'The Topographical List of the Brihat-Samhita', *Indian Antiquary*.
Furer-Haimendorf, Christoph Von (1982): *Tribes of India: The Struggle for Survival*, New Delhi: Oxford University Press.
Ghurye, G.S. (1950): *Caste and Class in India*, Bombay: Popular Prakashan.
__________(1963): *The Scheduled Tribes*, Bombay: Popular Prakashan.
__________(1969): *Caste and Race in India*, Bombay: Popular Prakashan.

Gibb, H.A.R. (1929): *Travels of Ibn Batuta*, London.

__________(1962): *Travels of Ibn Batuta*, New Edition, Cambridge.

Government of India (1978): *Provisions in the Constitution of India for Scheduled Tribes*, Background Papers on Tribal Development, New Delhi: Ministry of Home Affairs.

Government of India (1978): *Report of the Study Team on Implementation of Tribal Development Programmes During the Medium Term Plan 1978-83*, New Delhi: Ministry of Home Affairs.

Government of India (1978): *Scheduled Tribes and Scheduled Areas in India*, Background Papers on Tribal Development, New Delhi: Ministry of Home Affairs.

Greenough, W.B. and R.A. Cash (1973): *Post-Civil War in Bangladesh*, New York: Oxford University Press.

Grierson, G.A. (1903): *The Languages of India*, Report on the Census of India 1901, Calcutta: Superintendent of Govt. Press.

Guha, B.S. (1944): *Racial Elements in the Population*, Bombay: Oxford University Press.

Gunatileke, G. (1973): 'The Rural-urban Balance and Development', *Marga*, 2 (1): 35-68.

Gupta, J.S. (1974): *History of Freedom Movement in Bangladesh*, 1942-1973, Some Involvement, Calcutta: Naya Prakash.

Gupta, S.L. (1963): 'Meghdoot—A Geographical Analysis', *National Geographical Journal of India*, 9 (3-4).

Gurung, H. (1968): 'Geographic Foundations of Nepal', *The Himalayan Review*, Special Issue 1: 1-9.

_________(1969): *Regional Development Planning for Nepal*, Kathmandu: National Planning Commission.

_________(1971): 'Demographic Aspects of Development in Nepal', *Population and Development*, Kathmandu: Centre for Economic Development and Administration.

Habib, Irfan (1963): The Agrarian System of Mughal India, Bombay.

__________(1982): *An Atlas of the Mughal Empire*, Oxford: Oxford University Press.

Hageen, T. (1961): *Nepal—The Kingdom in the Himalayas*, Bern: Kummerly.

Haig, M.R. (1887): 'Ibn Batuta's Travels in Sind', *Journal of the Royal Asiatic Society*.

Hartshorne, R. (1939): *The Nature of Geography*, Lancaster: Annals of the Association of American Geographers.

Hasan, Z. (ed.; 1918): *Khulasat-u-Tarikh of Suian Rai Bhandari*, Delhi.

Hasnat, A. (1974): *The Ugliest Genocide in History*, Dacca: Muktadhara.

Helbock, R.W. (1975): 'Differential Urban Growth and Distance Consideration in Domestic Migration Flows in Pakistan', *The Pakistan Development Review*, 14 (1): 53-84.

Husain, A. (1970): *British India's Relation with the Kingdom of Nepal: 1857-1947*, London: George Allen.

Hussain, C.A. (1980): 'Settlement of Refugees', Paper presented at the IGU Symposium on Development and Redistribution of Population in South Asia.

Hutton, J.H. (1963): *Caste in India: Its Nature, Function and Origin*, Bombay: Oxford University Press.

ILO. (1971): 'Matching Employment Opportunities and Expectations: A Programme of Action for Ceylon', Technical Papers, Geneva: International Labour Organisation.

Itagi, N.H. (1994): 'Spatial Variation in the Incidence of Bilingualism in the Tribal Population of India', in N.H. Itagi (ed.), *Spatial Aspects of Language*, pp. 175-85, Mysore: Central Institute of Indian Languages.

Jarrett, H.S. (ed.; 1891): English Translation of *Ain-e-Akbari*, 2, from 1884-89, Annotated by Sir Jadu-nath Sarkar, Calcutta: Royal Asiatic Society of Bengal, 1949.

Jathar, G.B. and Beri, S.G. (1949): *Indian Economics—A Comprehensive and Critical Survey*, London: Oxford University Press.

Jillani, M.S. (1963): 'Resettlement Pattern of Displaced Persons in Pakistan', *Geographia*, 2 (2): 77-98.

Johnson, B.L. (1979): *Pakistan*, London: Heinemann.

Jones, G. and S. Selvaratham (1970): 'Urbanisation in Ceylon, 1946-63', *Modern Ceylon Studies*, 1 (2).

Joppen, S.J. Charles (1917): *Historical Atlas of India*, London: Longmans, Green and Co.

Joshi, P.C. (1976): *Land Reforms in India—Trends and Perspectives*, New Delhi: Allied Publishers Limited.

Kamaluddin, A.F.M. (1985): 'Refugee Problem in Bangladesh', in Leszek A. Kosinski and K. Maudood Elahi (ed.), *Population Redistribution and Development in South Asia*, pp. 221-36, Dordrecht: D. Reidel Publishing Company.

Kansakar, V.B.S. (1973-74): 'History of Population Migration in Nepal', *The Himalayan Review*, 6: 58-68.

____________(1979): 'Effectiveness of Planned Resettlement Programme in Nepal', Paper No. 1, Kathmandu: Centre for Economic Development and Administration.

Karan, P.P. (1967): *Bhutan—A Physical and Cultural Geography*. Lexington: University of Kentucky Press.

Karimi, S.M. (1970): 'Akbarnama', *The Geographer*, Aligarh.

Kashyap, S.C. (1971): 'Bangladesh—An Introduction and Study in Background', in S.C. Kashyap (ed.), *Bangladesh: Background and Perspective*, Delhi: National Publishers.

Khaikar, V.P. (2003): 'Migration in the City of Pune—A Case Study', *Annals of the National Association of Geographers, India*, 23(1): 84-94.

Khan, Hashmatullah (1939): *Mukhtasar-Tarikh-I-Jammu*, Lucknow.

Khan, I.R. (1934): 'Historical Geography of the Punjab and the Sind with Special Reference to River Changes', *The Aligarh Muslim University Journal*, 2(1).

Khan, M.A. (1981): *Spatial Analysis of Demographic Patterns and Trends in North West Frontier Province*, Peshawar: Institute of Economic Studies.

Khan, Rasheeduddin (1977): 'The Changing Role of Religious Minorities' in Romesh Thapar (ed.), *Tribe, Caste and Religion*, pp. 143-60, New Delhi: Macmillan Company of India Ltd.

Khan, Rasheeduddin (1992): *Federal India—A Design for change*, New Delhi: Vikas Publishing House.

____________(ed.; 1997): *Rethinking Indian Federalism*, Delhi: Raplica Press Pvt. Ltd.

Khubchandani, M.L. (1983): *Plural Languages, Plural Cultures: Communication, Identity and Socio-political Change in Contemporary India*, Honolulu: East-West Center.

____________(1988) *Language in a Plural Society*, Delhi: Motilal Banarasidas.

____________(1993): 'India as a Socio-linguistic Area', in Aijazuddin Ahmad (ed.), *Social Structure and Regional Development: A Social Geography Perspective*, pp. 101-13, Jaipur: Rawat Publishers.

____________(1994): 'India as a Socio-linguistic Area', in N.H. Itagi (ed.), *Spatial Aspects of Language*, pp. 33-43, Mysore: Central Institute of Indian Languages.

Knight, D.B. (1982): 'Identity and Territory: Geographical Perspectives on Nationalism and Regionalism', *Annals of the Association of American Geographers*, 72: 514-531.

Kosambi, D.D. (1962): *Myth and Reality—Studies in the Formation of Indian Culture*, Bombay: Popular Prakashan.

____________(1972): *The Culture and Civilisation of Ancient India—An Historical Outline*, Delhi: Vikas Publishing House Pvt. Ltd.

Kosinski, Leszek A. and K. Maudood Elahi (eds.; 1985): *Population Redistribution and Development in South Asia*, Dordrecht: D. Reidel Publishing Company.

Krishna, Gopal (1977): 'Framework of Minority Politics', in Romesh Thapar (ed.), *Tribe, Caste and Religion*, pp. 161-68, New Delhi: The Macmillan Company of India Ltd.

Krishan, Gopal and Neela Thapar (1994): 'Cultural Diversity and Development in India', *Annals of the National Association of Geographers, India*, 14(2): 27-44.

Krishnan, Gopal (1995): 'Demographic Space Expansion by Minorities in India', *Annals of the National Association of Geographers, India*, 15(2): 7-25.

Krishnaswami, V.D. (1956): 'The Neolithic Pattern in India', Presidential Address of the Anthropology Section of the Indian Science Congress.

Krotki, K.J. (1976): 'Pakistan's Population Size and Growth in the Light of 1972 Census: Evaluation Survey', *The Pakistan Development Review*, 15(2): 181-94.

Kuznets, Simon and D.S. Thomas (1960): *Population Redistribution and Economic Growth: United States, 1870-1950*, Philadelphia: American Philosophical Society.

Lall, K. (1968): 'The Nepalese without Nepal', *Nepal Review*, 1(8): 343-46.

Landstrom, Bjorn (1964): *The Quest for India*, Stockholm: Doubleday.

Law, B.C. (1954): *Geographical Aspect of Kalidas's Works*, Calcutta.

________(1955): 'Al-Beruni's Knowledge of Indian Geography', *Indo Iranica*, 7(4).

Mahbub-ul-Haq (2002): *Human Development in South Asia*, Human Development Centre: Oxford.

Majumdar, R.C. (ed.; 1927): *McCrindle's Ancient India as Described by Ptolemy*, Calcutta: ASI.

____________(1960): *Classical Accounts of India*, Calcutta: ASI.

Majumdar, S.N. (1919-21): 'Contributions to the Study of the Ancient Geography of India', *Indian Antiquary*.

Mandelbaum, D.G. (1968): 'Family, Jati, Village', in B.S. Cohn and Milton Singer (eds.), *Structure and Change in Indian Society*, pp. 29-31, Chicago: Aldine Publishing Company.

______________(1970): *Society in India*, Bombay: Popular Prakashan.

Martine, G. (1979): 'Adaptation of Migrants or Survival of the Fittest: A Brazilian Case', *The Journal of Developing Areas*, 14 (1): 23-42.

McCrindle, J. W. (1885): *Ancient India as Described by Ptolemy*, Calcutta.

____________(1926): *Ancient India as Described by Megasthenes and Arrien.*

Mikesell, Marvin W. (ed.; 1973): 'Geographers Abroad: Essays on the Problems and Prospects of Research in Foreign Areas', Research Paper No. 152, Chicago: Department of Geography, University of Chicago.

Misra, R.P., K.V. Sundaram and V.L.S. Prakasa Rao (1978): *Regional Development Planning in India: A New Strategy*, New Delhi: Vikas.

Moonis Raza, Aijazuddin Ahmad and S.C. Nuna (1986): 'Tribal Literacy in India—The Regional Dimension', Occasional Paper No. 9, New Delhi: National Institute of Educational Planning and Administration.

Moreland, W.H. (1929): *The Agrarian System of Moslem India.*

Muhith, A.M.A. (1978): *Bangladesh—Emergence of a Nation*, Dacca: Books International.

Mukerji, A.B. (1969): 'Rural Setlements in the Jatakas', *Geographical Review of India*, 31(2).

Myrdal, G. (1968): *Asian Drama*, New York: Penguin Books.

Nawaz Dawood (1980): Tea and Poverty: Plantations and the Political Economy of Sri Lanka, Urban Rural Mission—Hong Kong: Christian Conference of Asia.

Nehru, Jawaharlal (1962): *Jawaharlal Nehru—An Autobiography*, New Delhi: Allied Publishers Pvt. Ltd.

Nesfield, J.C. (1855): *Brief View of the Caste System of the North Western Provinces and Oudh.*

Omvedt, Gail (ed.; 1982): *Land, Caste and Politics in Indian State*, Delhi: Authors Guild Publications.

Oyugi, Walter O. (2000): 'Decentralisation for Good Governance and Development: Concepts and Issues', *Regional Development Dialogue*, 21(1): 3-22.

Panday, R.K. (1998): *Pioneers of Nepalese Geography—Geography and Geographers in Nepal*, Kathmandu: Centre for Nepal Geography.

Panikkar, K.M. (1955): *Geographical Factors in Indian History*, Bombay: Popular Prakashan.

Pargiter, F.E. (1894): 'Geography of Rama's Exile', *Journal of the Royal Asiatic Society of Bengal.*

Philot, D.C. (1927): *Ain-I-Akbari*, Calcutta: Royal Asiatic Society of Bengal.

Piggott, Stuart (1950): *Pre-Historic India*, Harmondsworth: Penguin Books.

Prabhu, P.H. (1956): 'Social Aspects of Urbanisation of Industrial Workers Migrating from Rural Areas to the City of Bombay', in *Social Implications of Industrialisation and Urbanisation*, Delhi: UNESCO Research Centre.

Puvanarajan, P. (1976): 'Patterns and Processes of Urbanisation in Sri Lanka', Population Problems of Sri Lanka, pp. 99-107, Colombo: Demographic Training and Research Centre Unit.

Rana, S.J.B. and Y.S. Thapa (1974): 'Population Migration: Causes, Consequences and Policy Implications', in C.C. Upadhayaya and J.V. Abueva (eds.), *Population and Development in Nepal*, pp. 79-96, Kathmandu: Centre for Economic Development and Administration.

Raverty, H.G. (1892): 'The Mehran of Sind and its Tributaries', *Journal of the Asiatic Society of Bengal*, 61 (3 and 4).

___________(1970): *Tabaqat-i-Nasiri*, Delhi.

Raychaudhuri, H.C. (1960): *Studies in Indian Antiquities*, Calcutta.

Raza, Moonis and Aijazuddin Ahmad (1990): *An Atlas of Tribal India*, New Delhi: Concept.

Risley, Sir H.H. (1891): *Tribes and Castes of Bengal*, Ethnographic Glossary.

___________(1915): *The People of India*, Oriental Books: Delhi.

Ross, E.D. (1895): *Tarikh-i-Rashidi*, London.

Ross, Harley and Haqqi (1918): *Haft iqlim of Amin Ahmad Razi*, Calcutta, 1918, 1927, 1939.

Roy Burman, B.K. (1972): 'Tribal Demography: A Preliminary Appraisal', in K.S. Singh (ed.), *Tribal Situation in India*, pp. 39-50, Shimla: Indian Institute of Advanced Study.

___________(1972): 'Distribution of Scheduled Tribes of India: An Explanatory Geo-Cultural Appraisal', *Economic and Socio-cultural Dimension of Regionalisation*, Census Centenary Monograph No. 7, pp. 483-538, New Delhi: Office of the Registrar General.

Ruhunage, L.K. (1979): *Migrations of Sri Lankans to the Middle Eastern Countries*, Colombo: Ministry of Plan Implementation.

Sachidananda (1972): 'The Tribal Situation in Bihar', in S.C. Dube (ed.), *Tribal Situation in India*, Simla: Institute of Advanced Study.

Sakharov, I.V. (1972): 'Ethno-linguistic Geography of India: Facts and Problems', in A. Chandra Sekhar (ed.), *Economic and Socio-cultural Dimensions of Regionalisation*, Census Centenary Monograph No. 7, pp. 387-426, New Delhi: Office of the Registrar General.

Sarkar, N.K. (1957): *The Demography of Ceylon*, Colombo: Ceylon Government Press.

Sastri, Nilakanta (1981): *A History of South India*, London: Oxford University Press.

Schwartzberg, J.E. (1967): 'Prolegomena to the Study of South Asian Regions and Regionalism', in Robert I. Crane (ed.), *Regions and Regionalism in South Asian Studies: An Exploratory Study*, pp. 89-111, Durham: Duke University.

_______________(1968): 'Caste Regions of the North Indian Plain', in B.S. Cohn and Milton Singer (eds.), *Structure and Change in Indian Society*, pp. 81-113, Chicago: Aldine Publishing Company.

_______________(ed.; 1978): *A Historical Atlas of South Asia*, Chicago: University of Chicago Press.

Sharma, A. K. (1989): 'Growth of Muslim Population in India—Some Policy Issues', *The Eastern Anthropologist*, 42 (4): 353-64.

Sharma, R.S. (ed.; 1974): *Indian Society—Historical Probings*, New Delhi: Peoples Publishing House.

Shrestha, C.B. (1975): 'Urbanisation Trends and Emerging Pattern in Nepal', *The Himalayan Review*, 3(7): 1-13.

___________(1978): 'Vertical Distribution of Population in Nepal', *Journal of Institute of Humanities and Social Sciences*, 1: 37-45.

Singh, K.S. (1972): *Tribal Situation in India*, Shimla: Indian Institute of Advanced Study.

_________(1978): 'Colonial Transformation of Tribal Societies in Middle India', *Economic and Political Weekly*, 13(30): 1221-32.

_________(1985): *Tribal Society in India: An Anthropo-Historical Perspective*, New Delhi: Manohar Publications.

_________(1992) *People of India—An Introduction*, Calcutta: Anthropological Survey of India.

Singh, R.L. and R.B. Singh (1972): 'Spatial Diffusion of Rajput Clan Settlements in a Part of Middle Ganga Valley', in R.L. Singh (ed.), *Rural Settlements in Monsoon Asia*, pp. 152-70, Varanasi: National Geographical Society of India.

Singh, R.A. (1994): 'Historical Atlas of Indian' Languages—A Case Study of Migration of Indo-Aryan through the Ages in N.H. Itagi (ed.) *Spatial Aspects of Language*, pp. 187-95, Mysore: Central Institute of Indian Languages.

Sinha, Surajit (1982): 'Rethinking about Tribes and Indian Civilisation', in Budhadeb Choudhuri (ed.), *Tribal Development in India: Problems and Prospects*, pp. 3-14, New Delhi: Inter India Publications.

Sircar, D.C. (1960): *Studies in the Geography of Ancient and Medieval India*, Delhi.

Sivaramamurthy, C. (1932): 'Geography of Kalidasa', *Journal of Madras Geographical Association*, 7(1).

Sopher, D.E. (1973): 'Towards a Rediscovery of India: Thoughts on Some Neglected Geography', in M.M. Mikesell (ed.), *Geographers Abroad: Essays on the Problems and Prospects of Research in Foreign Areas*, Paper No. 153, pp. 110-32, Chicago: Department of Geography, University of Chicago.

_________(1974): 'Indian Pastoral Castes and Livestock Ecologies: A Geographical Analysis', in L.S. Leshnik and G.D. Sontheimer (eds.), *Nomads and Pastoralists in South Asia*, Wiesbaden.

_________(1977): 'Rohilkhand and Oudh: An Exploration of Social Gradients across a Political Frontier', in Richard C. Fox (ed.), *Realm and Region in Traditional India*, N.C. Chapel Hill.

_________(1987): 'The Message of Place in Hindu Pilgrimage', *The National Geographical Journal of India*, 33(4): 353-69.

Spencer, G.W. (1969): 'Religious Network and Royal Influence in Eleventh Century South India', *Journal of the Economic and Social History of the Orient*, 12: 42-56.

Sri Lanka Department of Labour and ILO (1972): *Report of the Seminar on Population and Family Planning*, Colombo.

Srinivas, M.N. (1962): *Caste in Modern India and Other Essays*, Bombay: Asia Publishing House.

__________(1986): *India: Social Structure*, Delhi: Hindustan Publishing Corporation.

__________(1992): 'On the People of India Project', Foreword to K.S. Singh, *People of India—An Introduction*, pp. 1-12, Calcutta: Anthropological Survey of India.

Stein, A. (1900): *Kalhana's Rajatarangini*, 1 and 2, London.

Stein, M.A. (1971): 'Some River Names in the Rigveda', *Journal of the Royal Asiatic Society*.

Subbarao, B. (1958): *The Personality of India*. Baroda: M.S. University.

Sukhwal, B.L. (1985): *Modern Political Geography of India*, New Delhi: Sterling Publishers.

Tamaskar, B.G. (1956): 'Geographical Knowledge in the Upanishads', *National Geographical Journal of India*, 2(2).

Tamaskar, B.G. (1967-68): 'Geographical data in the Jataka Tales', *Quarterly Review of Historical Society*, 12(2).

Tehran University (1971): *Historical Atlas of Iran*, Tehran: Tehran University Press.

Thapa, Y.S. and P.N. Tiwari (1977): *Immigration Pattern in Kathmandu Urban Areas*, Kathmandu: Centre for Economic Development and Administration.

Thapar, Romila (1977): *A History of India 1*, Harmondsworth: Penguin Books.

__________(1990): *From Lineage to State*, New Delhi: Oxford University Press.

Tod, J. (1829): *Annals and Antiquities of Rajasthan*, 1-2, London.

Todaro, M.P. (1976): *Internal Migration in Developing Countries: A Review of Theory, Evidence, Methodology and Research Priorities*, Geneva: International Labour Organisation.

__________(1977): *Economic Development in the World: An Introduction to Problems and Policies in Global Perspective*, London: Longman.

Toynbee, Arnold J. (1961): *Between Oxus and Jumna*, New Delhi: Oxford University Press.

Turner, Frederick Jackson (1894): 'The Significance of the Frontier in American History', Annual Report of the American Historical Association, 1893, Washington D.C.: U.S. Government Printing Press.

United Nations High Commission for Refugees (1981): 'Humanitarian Assistance Programme to Afghan Refugees in Pakistan', Report No. 9.

Verma, K.K. (1977): *Culture, Ecology and Population: An Anthropo-demographic Study*, New Delhi: National Publishing House.

Wadia, D.N. (1976): *Geology of India*, New Delhi: Tata McGraw-Hill Publishing Company Ltd.

Wheeler, R. Mortimer (1959): *Early India and Pakistan*, London.

Wilford, F. (1822): 'On the Ancient Geography of India', *Asiatic Researchers*, 14.

World Commission on Environment and Development (1987): *Our Common Future*, Oxford: Oxford University Press.

World Health Organisation (2000): *Global Water Supply and Sanitation Assessment 2000 Report*, WHO and United Nations Children's Fund.

Yule, H. (1874): 'Geography of Ibn Batuta's Travels in India', *Indian Antiquary*.

Zelinsky, W. (1971): 'The Hypothesis of Mobility Transition', *Geographical Review*, 61(2): 219-49.

Index